MUSCULAR SYSTEM COLORING BOOK

GUIDE FOR THE MUSCULAR SYSTEM

*New for 2016

Now you can learn and master the muscular system with ease while having fun reinforcing visual details of the muscular system.

By: Pamphlet Books

MUSCULAR SYSTEM

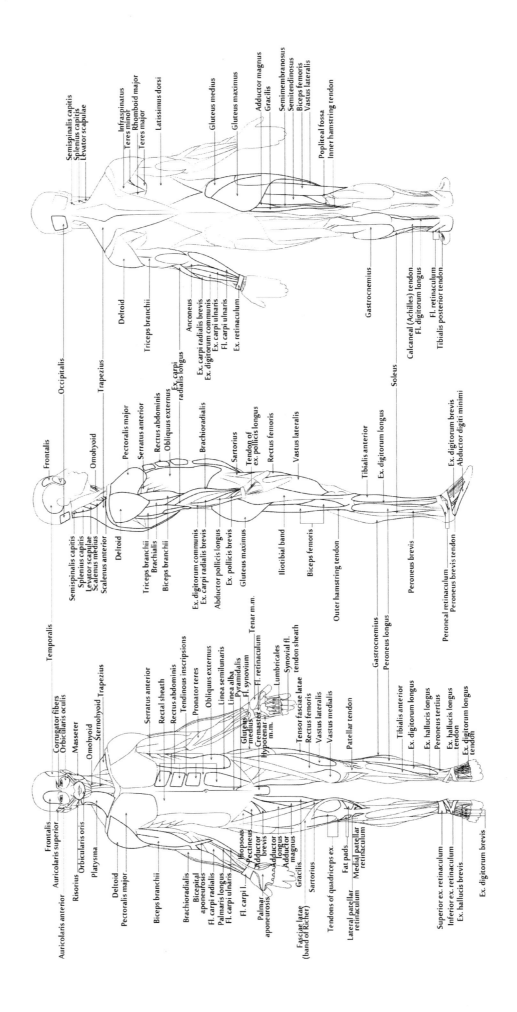

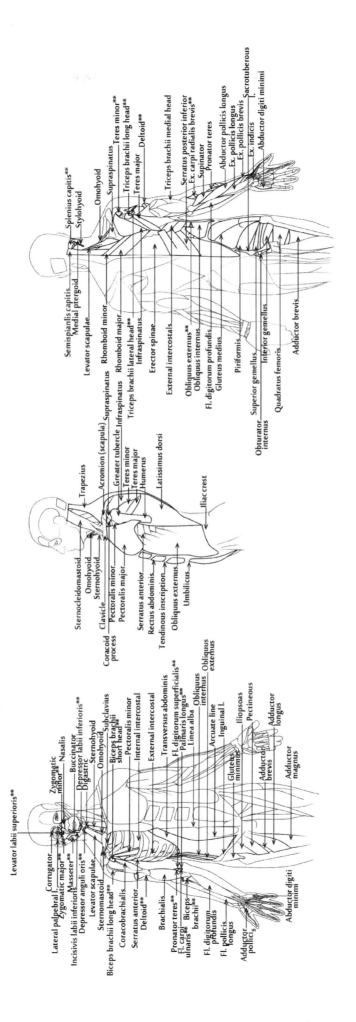

DEEP MUSCLE

HEAD MUSCLE

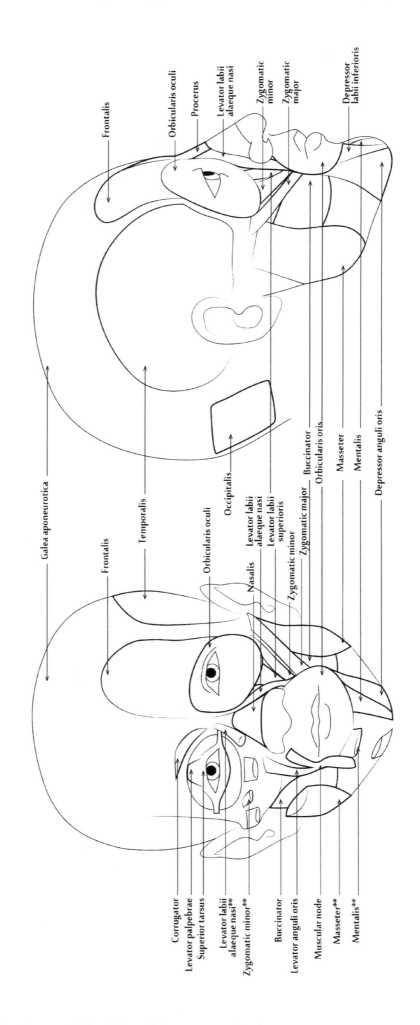

Galea aponeurotica

Frontalis

Temporalis

Orbicularis oculi

Occipitalis

Nasalis

Levator labii alaeque nasi
Levator labii superioris

Zygomatic minor

Zygomatic major

Buccinator
Orbicularis oris

Masseter

Mentalis

Depressor anguli oris

Frontalis

Orbicularis oculi

Procerus

Levator labii alaeque nasi

Zygomatic minor

Zygomatic major

Depressor labii inferioris

Corrugator
Levator palpebrae
Superior tarsus
Levator labii alaeque nasi**
Zygomatic minor**
Buccinator
Levator anguli oris
Muscular node
Masseter**
Mentalis**

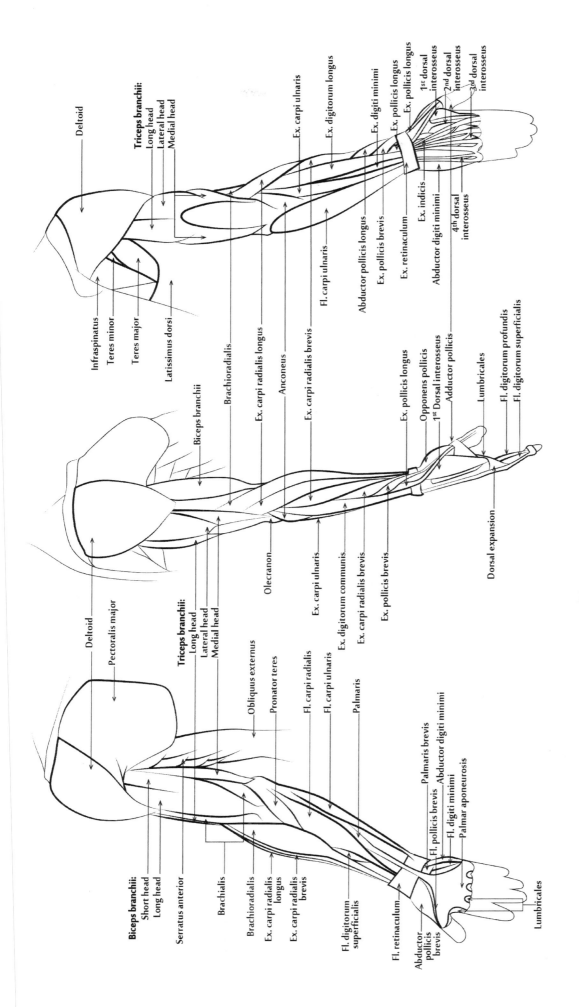

ARM MUSCLE

PAGES 32-41

LEG MUSCLE

Gluteus medius

Gluteus maximus

Adductor magnus

Gracilis

Semimembranosus

Biceps femoris:
Long head
Short head

Fibular head

Gastrocnemius lateral head

Soleus

Peroneus longus

Calcaneal (Achilles) tendon

Fl. hallucis longus

Iliotibial band

Semitendinosus

Vastus lateralis

Plantaris

Polipteal fossa

Gastrocnemius:
Lateral head
Medial head

Sartorius

Soleus

Peroneus brevis

Fl. hallucis longus tendon

Peroneus longus

Peroneal retinaculum

Fl. digitorum longus tendon

Tibialis posterior tendon

Calcaneus

Tibialis posterior tendon

Gluteus medius

Sartorius

Tensor fasciae latae

Rectus femoris

Iliotibial band

Vastus intermedius

Vastus lateralis

Patella

Patellar l.

Tibialis anterior

Peroneus tertius

Ex. retinaculum

Ex. digitorum brevis

Abductor digiti minimi

Peroneus tertius

Iliopsoas

Pectineus

Adductor brevis

Adductor longus

Adductor magnus

Gracilis

Vastus medialis

Semitendinosus

Medial meniscus

Gastrocnemius medial head

Soleus

Ex. hallucis longus

Ex. retinaculum

Ex. hallucis brevis

Gluteus medius

Tensor fasciae latae

Iliotibial band

Rectus femoris

Sartorius

Vastus lateralis

Patella

Lateral meniscus

Patellar l.

Peroneus longus

Gastrocnemius lateral head

Ex. digitorum longus

Peroneus brevis

Peroneus tertius

Ex. digitorum brevis

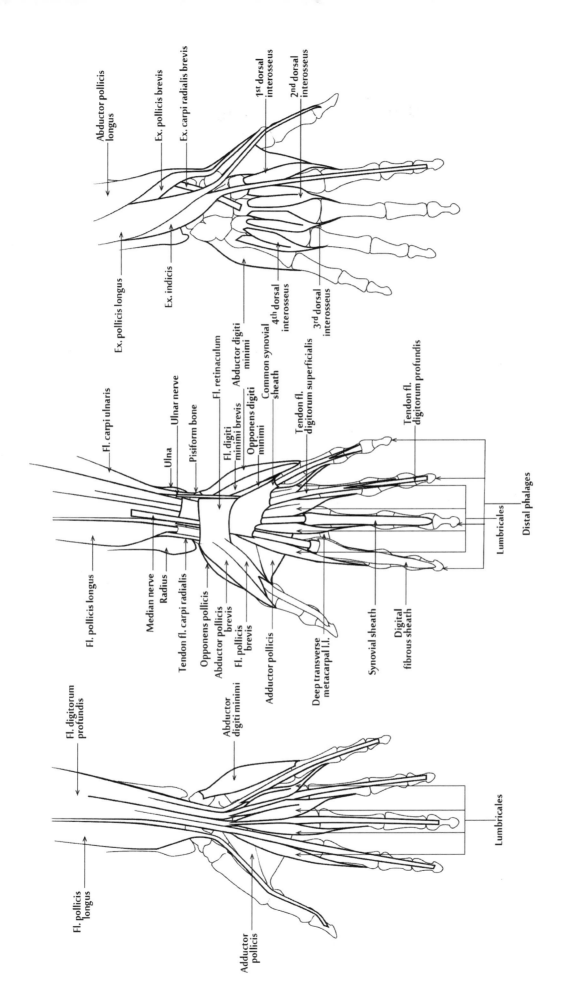

HAND MUSCLE

Abductor pollicis longus

Ex. pollicis brevis

Ex. carpi radialis brevis

1st dorsal interosseus

2nd dorsal interosseus

Ex. pollicis longus

Ex. indicis

4th dorsal interosseus

3rd dorsal interosseus

Fl. carpi ulnaris

Ulnar nerve

Ulna

Pisiform bone

Fl. retinaculum

Fl. digiti minimi brevis

Abductor digiti minimi

Opponens digiti minimi

Common synovial sheath

Tendon fl. digitorum superficialis

Tendon fl. digitorum profundis

Median nerve

Radius

Tendon fl. carpi radialis

Opponens pollicis

Abductor pollicis brevis

Fl. pollicis brevis

Adductor pollicis

Deep transverse metacarpal l.l.

Synovial sheath

Digital fibrous sheath

Lumbricales

Distal phalages

Fl. pollicis longus

Fl. digitorum profundis

Abductor digiti minimi

Fl. pollicis longus

Adductor pollicis

Lumbricales

FOOT MUSCLE

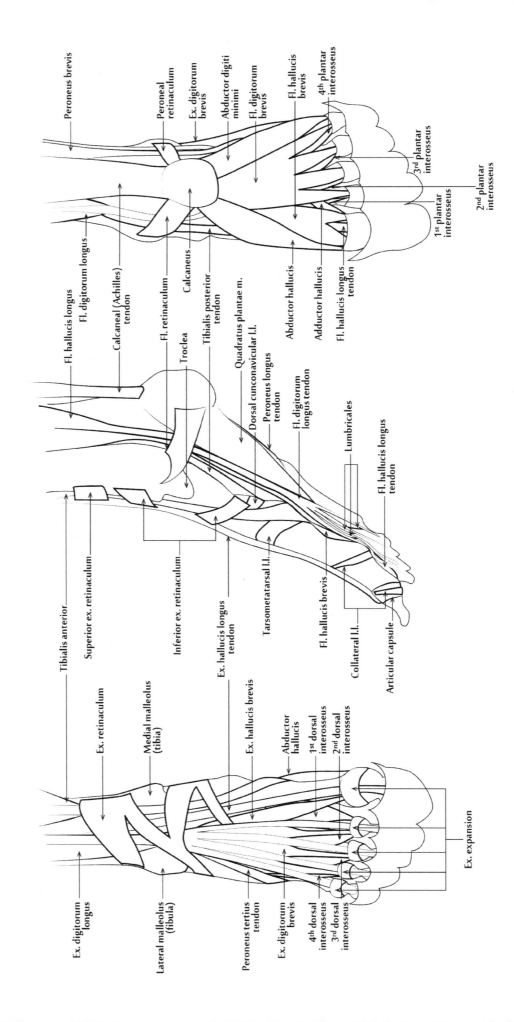

Peroneus brevis

Peroneal retinaculum

Ex. digitorum brevis

Abductor digiti minimi

Fl. digitorum brevis

Fl. hallucis brevis

4th plantar interosseus

3rd plantar interosseus

2nd plantar interosseus

1st plantar interosseus

Abductor hallucis

Adductor hallucis

Fl. hallucis longus tendon

Fl. hallucis longus

Fl. digitorum longus

Calcaneal (Achilles) tendon

Troclea

Fl. retinaculum

Calcaneus

Tibialis posterior tendon

Quadratus plantae m.

Dorsal cunconavicular l.l.

Peroneus longus tendon

Fl. digitorum longus tendon

Lumbricales

Fl. hallucis longus tendon

Tibialis anterior

Superior ex. retinaculum

Inferior ex. retinaculum

Ex. hallucis longus tendon

Tarsometatarsal l.l.

Fl. hallucis brevis

Collateral l.l.

Articular capsule

Ex. digitorum longus

Ex. retinaculum

Medial malleolus (tibia)

Ex. hallucis brevis

Abductor hallucis

1st dorsal interosseus

2nd dorsal interosseus

Lateral malleolus (fibula)

Peroneus tertius tendon

Ex. digitorum brevis

4th dorsal interosseus

3rd dorsal interosseus

Ex. expansion

PUBLISHERS NOTES

Disclaimer

This publication is intended to provide helpful and informative material.

The author and publisher specifically disclaim all responsibility for any liability, loss or risk, personal or otherwise, which is incurred as a consequence, directly or indirectly, from the use or application of any contents of this coloring work book.

Any and all product names referenced within this coloring work book are the trademarks of their respective owners. None of these owners have sponsored, authorized, endorsed, or approved this coloring work book.

Paperback Edition

Manufactured in the United States of America

DEDICATION

I would like to dedicate this coloring book to all the dedicated people in the medical field.

STUDYING THE MUSCULAR SYSTEM

Studying and remembering the muscular system can be overwhelming!

The beautifully illustrated Muscular System Coloring Book are just what the doctor ordered for Medical Students, CNA's, EMT's, Paramedics, Medical technology, Nursing Students, Students of anatomy, Psychology, Nurses, Sports Trainers, Specialists, Educators, Biology, Fitness education, Practitioners, Chiropractors, Reflexologist, Researchers, Health administration, Therapists, Anatomists, Physiology, Injury Attorneys, and other Health Care providers.

The coloring work book is a unique study aid that gives students an innovative approach to learning. A priceless resource that will appeal to all students who are studying the muscular system, it's very detailed, well-designed, and precise.

The precise clear labeling of the anterior, lateral, posterior and deep muscles of the head, arm, hand, upper body, leg, and foot is a vital part of the learning process that's vital for any student learning the muscular system. You will be engaged while learning the muscular system.

Benefit from realistic illustrations that will help you master the muscular system with effortlessness while you're having fun coloring the different detailed muscles of the body and then comparing them with a labeled version .

Having a better understanding and learning the muscular system in detail can be achieved through coloring, coloring will improve your studying ability and help increase your reference recall by fixating the anatomical images in your mind for easy visual recall later on just from the simple physical activity of coloring.

The hold activity process of coloring is intended to imprint on your memory the different shapes and location of each muscles, which will help you to visually recall later the different shapes and location of each muscle.

So instead of hours and hours and hours of memorization, the muscular system coloring book will help you learn through an interactive approach.

Unlabeled And Labeled Illustrations

How to use this book

1. You will need one box of coloring pencils.

2. You'll have an unlabeled and labeled example of each illustration to color along with a colored illustration that you can refer to for your coloring ideas.

3. To help you remember the location of each muscle use the labeled illustrations as your reference and color each muscle and find its label on the page to the left.

4. The physical act of doing this will reinforce the location of each muscle in your memory and help you remember each muscle name and its location.

5. You can also color the non-colored labeled illustration.

Cut = **
Extensor = ex.
Flexor = fl.
Ligament = l.
Ligaments = l.l.
Muscle = m.
Muscles = m.m.

Unlabeled Colouring Instructions

All labels are arranged in a random manner, each label is formatted in one line with its own font size, style and orientation. For example (in the label shown to your right) "DELTOID" is coloured yellow, and the muscle is filled with same colour. Please colour all muscles with different colours and find their labels (on the page to the left) and fill them in with the same colour as the muscle. If you run out of colours, start colouring by using two colour pencils for each label and muscle.

For example, Light Blue + Yellow = Green, hopefully different green from what is already in your colour set.

We would love to know what you think about this coloring book and how to make it better.

If you could take five minutes and go back to the site where you made your purchase and scroll to the bottom of that page and leave us a review.

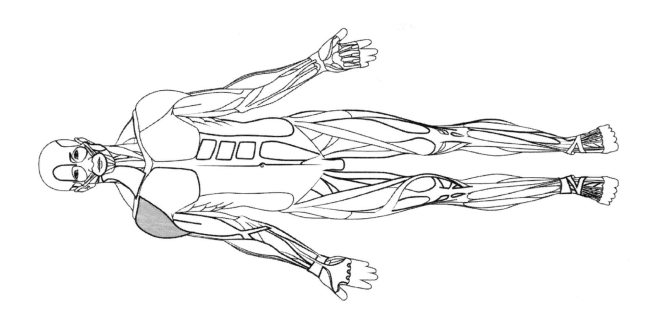

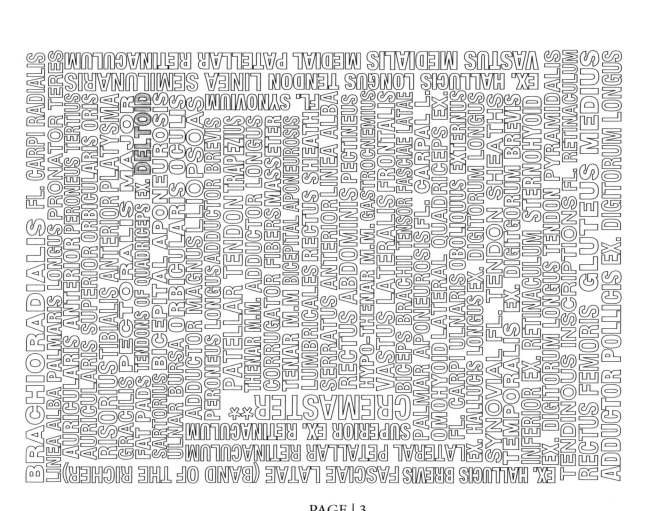

Notes

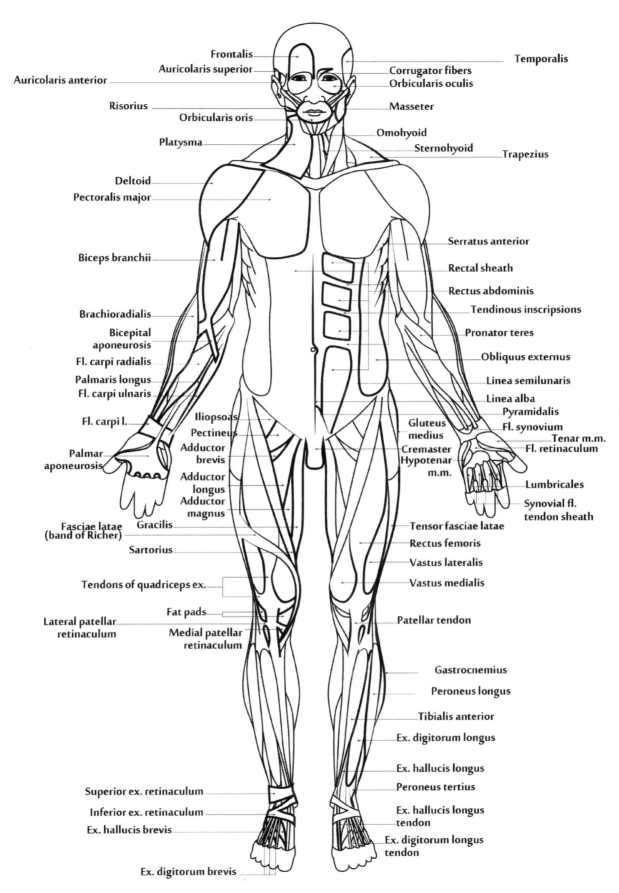

Frontalis
Auricolaris superior
Auricolaris anterior
Risorius
Orbicularis oris
Platysma
Deltoid
Pectoralis major
Biceps branchii
Brachioradialis
Bicepital aponeurosis
Fl. carpi radialis
Palmaris longus
Fl. carpi ulnaris
Fl. carpi l.
Palmar aponeurosis
Fasciae latae (band of Richer)
Gracilis
Sartorius
Tendons of quadriceps ex.
Lateral patellar retinaculum
Fat pads
Medial patellar retinaculum
Superior ex. retinaculum
Inferior ex. retinaculum
Ex. hallucis brevis
Ex. digitorum brevis

Temporalis
Corrugator fibers
Orbicularis oculis
Masseter
Omohyoid
Sternohyoid
Trapezius
Serratus anterior
Rectal sheath
Rectus abdominis
Tendinous inscripsions
Pronator teres
Obliquus externus
Linea semilunaris
Linea alba
Pyramidalis
Fl. synovium
Tenar m.m.
Fl. retinaculum
Lumbricales
Synovial fl. tendon sheath
Tensor fasciae latae
Rectus femoris
Vastus lateralis
Vastus medialis
Patellar tendon
Gastrocnemius
Peroneus longus
Tibialis anterior
Ex. digitorum longus
Ex. hallucis longus
Peroneus tertius
Ex. hallucis longus tendon
Ex. digitorum longus tendon

Iliopsoas
Pectineus
Adductor brevis
Adductor longus
Adductor magnus

Gluteus medius
Cremaster
Hypotenar m.m.

Body Anterior Muscles Labeled

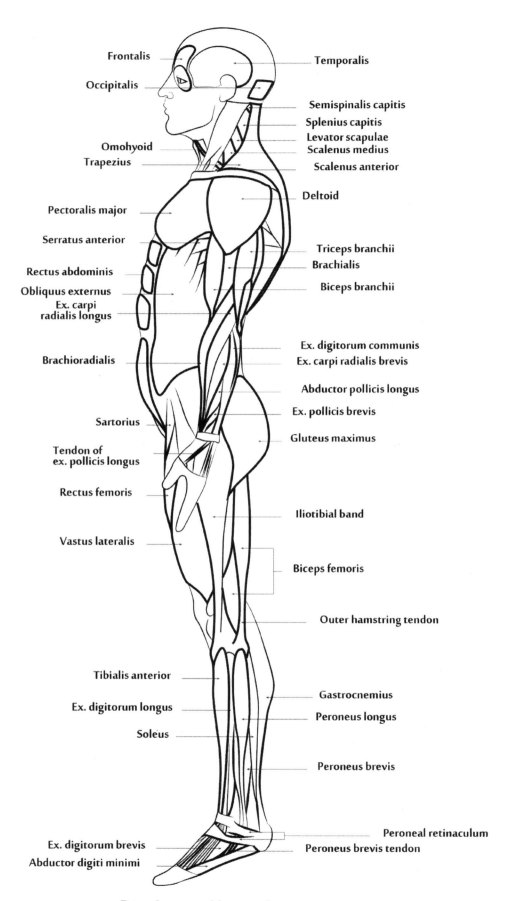

Frontalis

Occipitalis

Omohyoid

Trapezius

Pectoralis major

Serratus anterior

Rectus abdominis

Obliquus externus

Ex. carpi
radialis longus

Brachioradialis

Sartorius

Tendon of
ex. pollicis longus

Rectus femoris

Vastus lateralis

Tibialis anterior

Ex. digitorum longus

Soleus

Ex. digitorum brevis

Abductor digiti minimi

Temporalis

Semispinalis capitis

Splenius capitis

Levator scapulae

Scalenus medius

Scalenus anterior

Deltoid

Triceps branchii

Brachialis

Biceps branchii

Ex. digitorum communis

Ex. carpi radialis brevis

Abductor pollicis longus

Ex. pollicis brevis

Gluteus maximus

Iliotibial band

Biceps femoris

Outer hamstring tendon

Gastrocnemius

Peroneus longus

Peroneus brevis

Peroneal retinaculum

Peroneus brevis tendon

BODY LATERAL MUSCLES LABELED

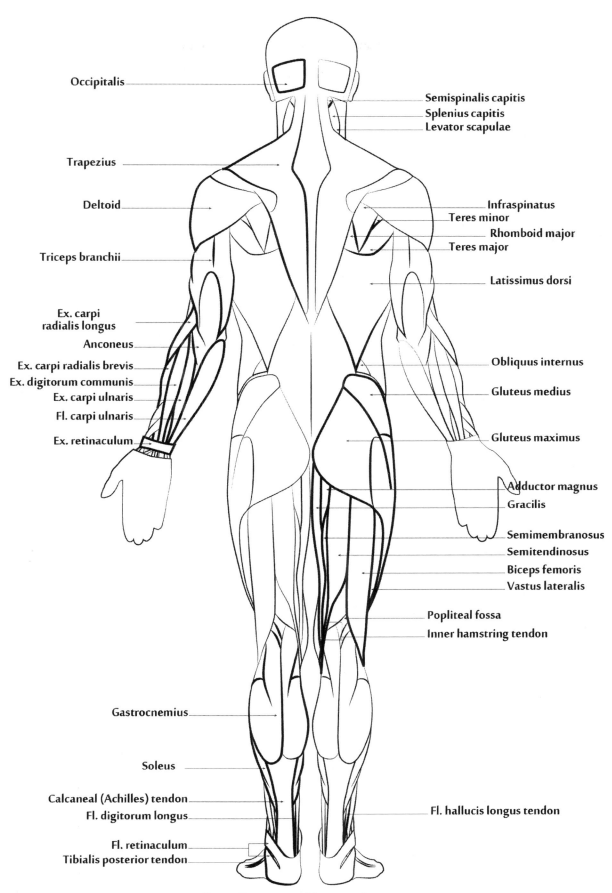

Occipitalis

Semispinalis capitis
Splenius capitis
Levator scapulae

Trapezius

Deltoid

Infraspinatus
Teres minor
Rhomboid major
Teres major

Triceps branchii

Latissimus dorsi

Ex. carpi
radialis longus

Anconeus

Ex. carpi radialis brevis
Ex. digitorum communis
Ex. carpi ulnaris
Fl. carpi ulnaris

Ex. retinaculum

Obliquus internus

Gluteus medius

Gluteus maximus

Adductor magnus
Gracilis

Semimembranosus
Semitendinosus
Biceps femoris
Vastus lateralis

Popliteal fossa
Inner hamstring tendon

Gastrocnemius

Soleus

Calcaneal (Achilles) tendon
Fl. digitorum longus

Fl. hallucis longus tendon

Fl. retinaculum
Tibialis posterior tendon

BODY POSTERIOR MUSCLES LABELED

BRACHIORADIALIS FL. CARPI RADIALIS
LINEA ALBA PALMARIS LONGUS PRONATOR TERES
AURICULARIS ANTERIOR PERONEUS TERTIUS
AURICULARIS SUPERIOR ORBICULARIS ORIS
RISORIUS TIBIALIS ANTERIOR PLATYSMA
GRACILIS PECTORALIS MAJOR
FAT PADS TENDONS OF QUADRICEPS EX. DELTOID
SARTORIUS BICEPITAL APONEUROSIS
ULNAR BURSA ORBICULARIS OCULIS
ADDUCTOR MAGNUS LLIOPSOAS
PERONEUS LONGUS ADDUCTOR BREVIS
PATELLAR TENDON TRAPEZIUS
THENAR M.M. ADDUCTOR LONGUS
CORRUGATOR FIBERS MASSETER
TENAR M.M BICEPITAL APONEUROSIS
LUMBRICALES RECTUS SHEATH
SERRATUS ANTERIOR LINEA ALBA
RECTUS ABDOMINIS PECTINEUS
HYPO-THENAR M.M. GASTROCNEMIUS
VASTUS LATERALIS FRONTALIS
BICEPS BRACHII TENSOR FASCIAE LATAE
PALMAR APONEUROSIS FL. CARPAL L.
OMOHYOID LATERAL QUADRICEPS EX.
FL. CARPI ULNARIS OBOLIQUUS EXTERNUS
EX. HALLUCIS LONGUS EX. DIGITORUM LONGUS
SYNOVIAL FL. TENDON SHEATHS
TEMPORALIS EX. DIGITGORUM BREVIS
INFERIOR EX. RETINACULUM STERNOHYOID
EX. DIGITORUM LONGUS TENDON PYRAMIDALIS
TENDINOUS INSCRIPTIONS FL. RETINACULUM
RECTUS FEMORIS GLUTEUS MEDIUS
ADDUCTOR POLLICIS EX. DIGITORUM LONGUS

EX. HALLUCIS BREVIS FASCIAE LATAE (BAND OF THE RICHER)
LATERAL PETALLAR RETINACULUM
SUPERIOR EX. RETINACULUM
CREMASTER**

FL. SYNOVIUM TENDON LINEA SEMILUNARIS MEDIAL PATELLAR RETINACULUM
EX. HALLUCIS LONGUS
VASTUS MEDIALIS

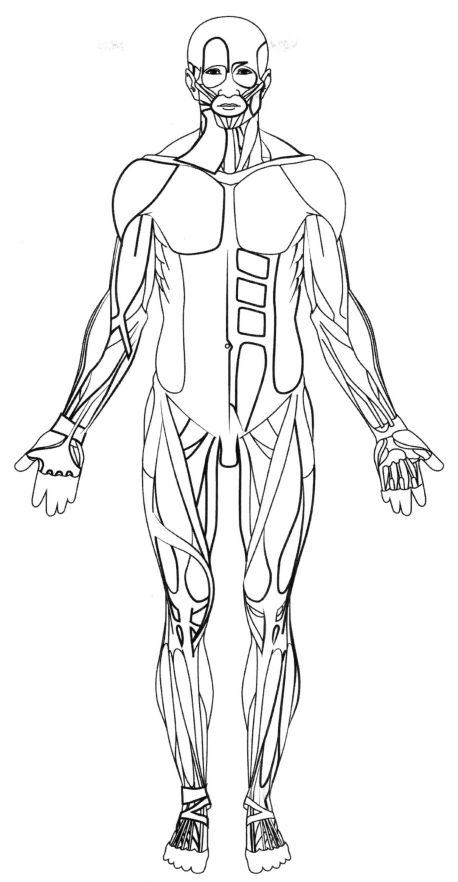

BODY ANTERIOR MUSCLE UNLABELED

SEMISPINALIS CAPITIS
SCALENUS ANTERIOR
EX. CARPI RADIALIS LONGUS
SPLENIUS CAPITIS TIBIALIS ANTERIOR
SCALENUS MEDIUS FRONTALIS
SERRATUS ANTERIOR TRAPEZIUS
SOLEUS PECTORALIS MAJOR
RECTUS ABDOMINIS OBLIQUUS EXTERNUS
BRACHIORADIALIS
RECTUS FEMORIS GASTROCNEMIUS
VASTUS LATERALIS BICEPS FEMORIS
EX. DIGITORUM LONGUS SARTORIUS
DELTOID TEMPORALIS
BRACHIALIS TRICEPS BRACHII
PERONEUS LONGUS BICEPS BRACHII
ILIOTIBIAL BAND GLUTEUS MAXIMUS
LEVATOR SCAPULAE
EX. CARPI RADIALIS BREVIS
OUTER HAMSTRING TENDON
OMOHYOID EX. POLLICIS BREVIS
EX. DIGITORUM BREVIS
EX. DITITORUM COMMUNIS
PERONEAL RETINACULUM
PERONEUS BREVIS TENDON

ABDUCTOR DIGITI MINIMI

POLLICIS LONGUS

TENDON OF EX.

OCCIPITALIS

PERONEUS BREVIS

ABDUCTOR POLLICIS LONGUS

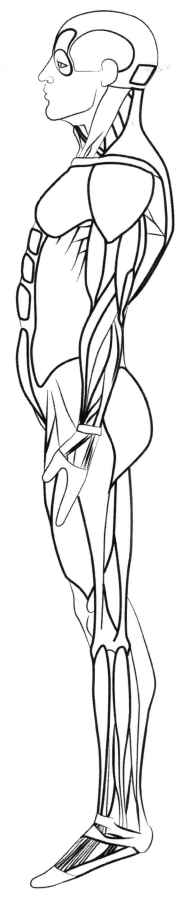

Body Lateral Muscles Unlabeled

BREVIS
TRICEPS BRACHII SOLEUS
SMIMEMBRANOSUS ANCONEUS
DELTOID TRAPEZIUS OCCIPITALIS
EX. CARPI RADIALIS LONGUS
EX. DIGITORUM COMMUNIS
EX. CARPI ULNARIS
FL. RETINACULUM
EX. RETINACULUM GRACILIS
FL. DIGITORUM LONGUS
TIBIALIS POSTERIOR TENDON
CALCANEAL (ACHILLES) TENDON
SEMISPINALIS CAPITIS
LATISSIMUS DORSI
OBLIQUUS INTERNUS
GLUTEUS MAXIMUS
LEVATOR SCAPULAE
INNER HAMSTRING TENDON
ADDUCTOR MAGNUS BICEPS FEMORIS
GLUTEUS MEDIUS VASTUS LATERALIS
SEMITENDINOSUS SPLENDIUS CAPITIS
FL. HALLUCIS LONGUS TENDON

EX. CARPI RADIALIS BREVIS
TERES MAJOR
TERES MINOR
RHOMBOID MAJOR
INFRASPINATUS
POPLITEAL FOSSA
GASTROCNEMIUS
FL. CARPI ULNARIS

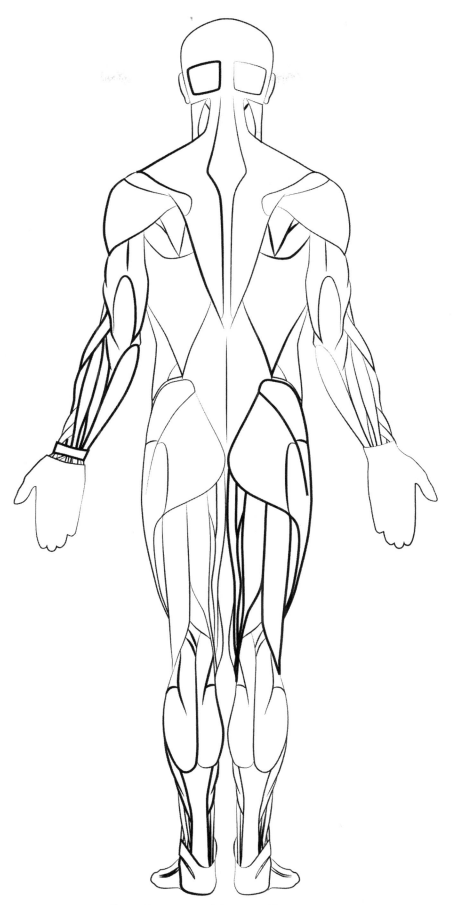

BODY POSTERIOR MUSCLES UNLABELED

NOTES

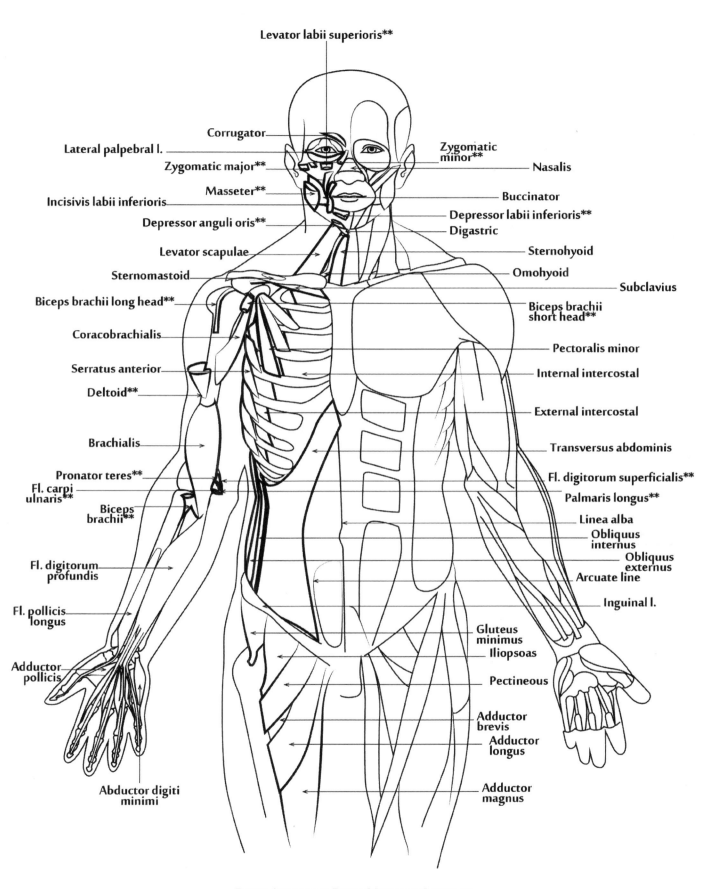

Levator labii superioris**

Corrugator

Lateral palpebral l.

Zygomatic major**

Masseter**

Incisivis labii inferioris

Depressor anguli oris**

Levator scapulae

Sternomastoid

Biceps brachii long head**

Coracobrachialis

Serratus anterior

Deltoid**

Brachialis

Pronator teres**

Fl. carpi ulnaris**

Biceps brachii**

Fl. digitorum profundis

Fl. pollicis longus

Adductor pollicis

Abductor digiti minimi

Zygomatic minor**

Nasalis

Buccinator

Depressor labii inferioris**

Digastric

Sternohyoid

Omohyoid

Subclavius

Biceps brachii short head**

Pectoralis minor

Internal intercostal

External intercostal

Transversus abdominis

Fl. digitorum superficialis**

Palmaris longus**

Linea alba

Obliquus internus

Obliquus externus

Arcuate line

Inguinal l.

Gluteus minimus

Iliopsoas

Pectineous

Adductor brevis

Adductor longus

Adductor magnus

BODY ANTERIOR DEEP MUSCLES LABELED

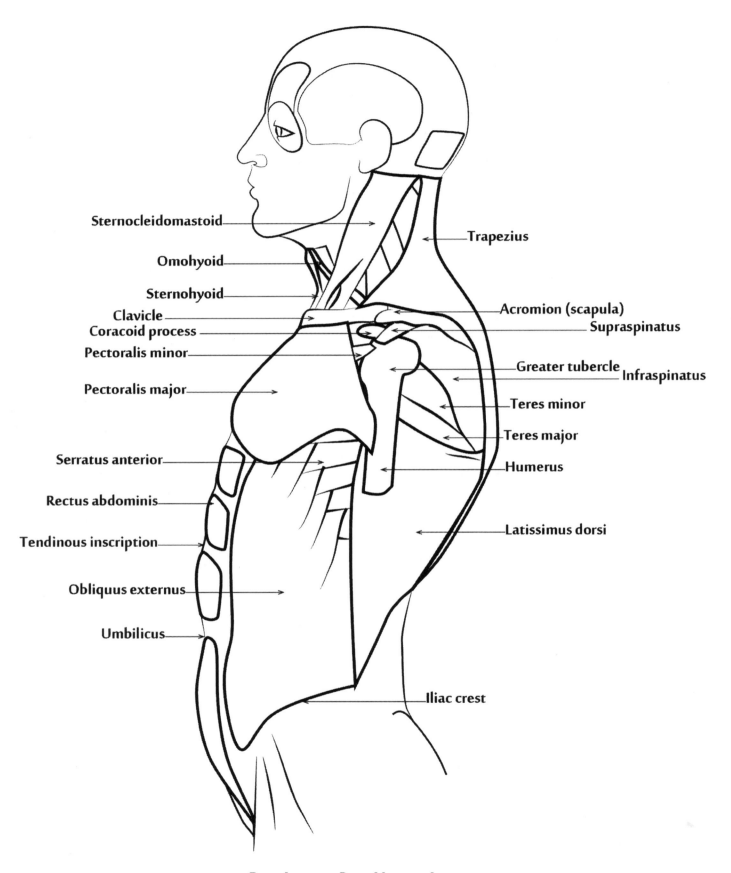

Sternocleidomastoid

Omohyoid

Sternohyoid

Clavicle

Coracoid process

Pectoralis minor

Pectoralis major

Serratus anterior

Rectus abdominis

Tendinous inscription

Obliquus externus

Umbilicus

Trapezius

Acromion (scapula)

Supraspinatus

Greater tubercle

Infraspinatus

Teres minor

Teres major

Humerus

Latissimus dorsi

Iliac crest

BODY LATERAL DEEP MUSCLES LABELED

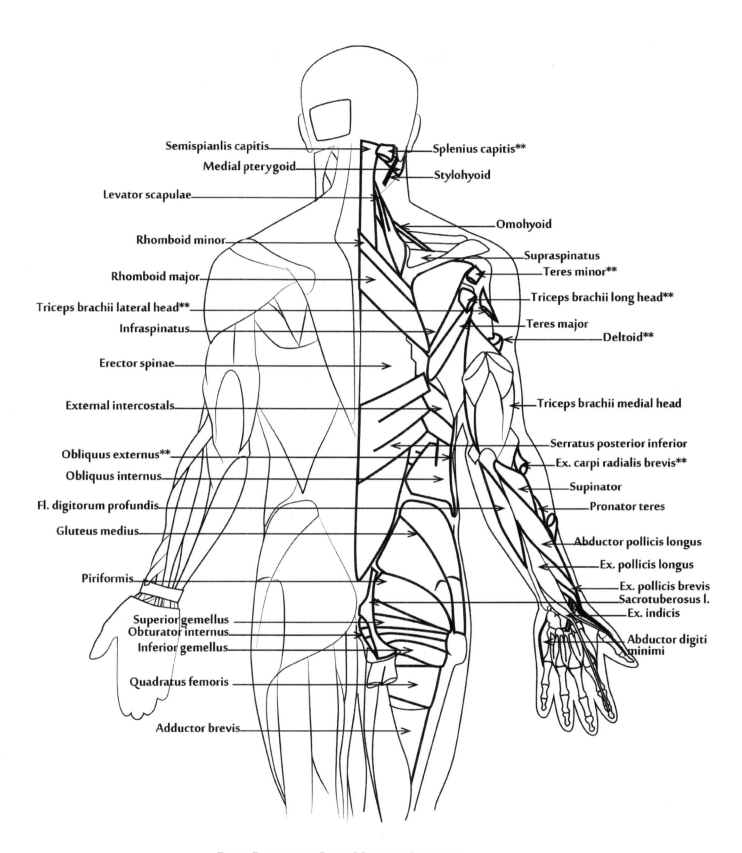

Semispianlis capitis

Medial pterygoid

Levator scapulae

Rhomboid minor

Rhomboid major

Triceps brachii lateral head**

Infraspinatus

Erector spinae

External intercostals

Obliquus externus**

Obliquus internus

Fl. digitorum profundis

Gluteus medius

Piriformis

Superior gemellus
Obturator internus
Inferior gemellus

Quadratus femoris

Adductor brevis

Splenius capitis**

Stylohyoid

Omohyoid

Supraspinatus

Teres minor**

Triceps brachii long head**

Teres major

Deltoid**

Triceps brachii medial head

Serratus posterior inferior

Ex. carpi radialis brevis**

Supinator

Pronator teres

Abductor pollicis longus

Ex. pollicis longus

Ex. pollicis brevis
Sacrotuberosus l.
Ex. indicis

Abductor digiti
minimi

BODY POSTERIOR DEEP MUSCLES LABELED

ADDUCTOR LONGUS INCISIVIS LABII INFERIORIS
LEVATOR LABII SUPERIORIS** BRACHIALIS
PRONATOR TERES** CORRUGATOR
BUCCINATOR LATERAL PALPEBRAL L.
SERRATUS ANTERIOR ADDUCTOR MAGNUS
CORACOBRACHIALIS ZYGOMATIC MAJOR**
BICEPS BRACHII**DEPRESSOR ANGULI ORIS**
ADDUCTOR BREVIS PECTORALIS MINOR
FL. CARPI ULNARIS**STERNOMASTOID**
BICEPS BRACHII LONG HEAD**
LEVATOR SCAPULAE DELTOID**
FL. POLLICIS LONGUS NASALIS
FL. DIGITORUM PROFUNDIS LINEA ALBA
ADDUCTOR POLLICIS**
ABDUCTOR DIGITI MINIMI
OMOHYOID STERNOHYOID
SUBCLAVIUS ILIOPSOAS
TRANSVERSUS ABDOMINIS
OBLIQUUS EXTERNUS**
MASSETER** PECTINEOUS
EXTERNAL INTERCOSTALS
INTERNAL INTERCOSTAL
OBLIQUUS INTERNUS**
ARCUATE LINE GLUTEUS MINIMUS
FL. DIGITORUM SUPERFICIALIS**

INGUINAL L. DIGASTRIC
PALMARIS LONGUS**

ZYGOMATIC MINOR**
BICEPS BRACHII SHORT HEAD**
DEPRESSOR LABII INFERIORIS**

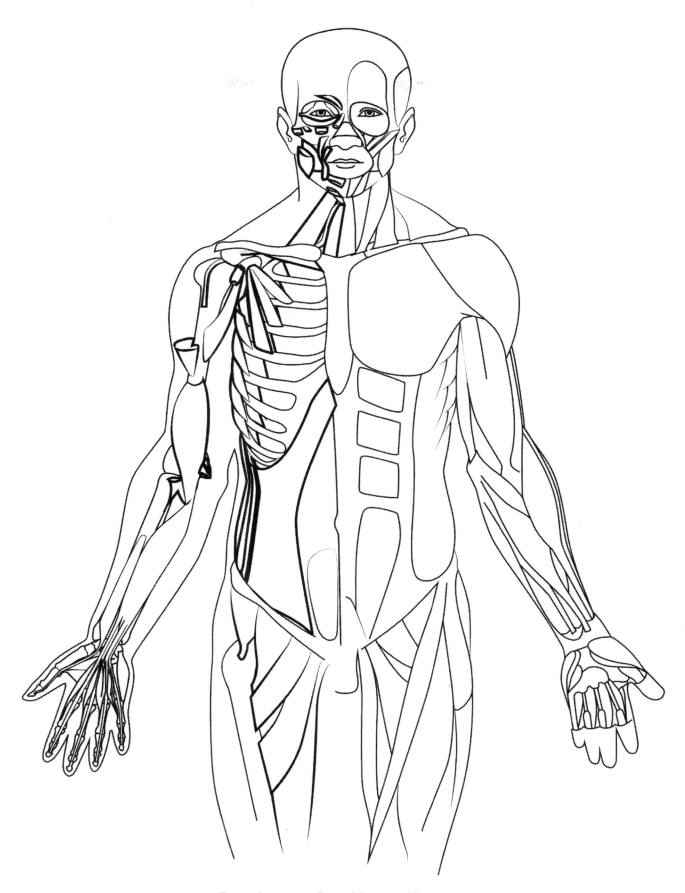

BODY ANTERIOR DEEP MUSCLES UNLABELED

EX. CARPI RADIALIS BREVIS
ABDUCTOR POLLICIS LONGUS
OBLIQUUS EXTERNUS
TRAPEZIUS SARTORIUS
TENDON OF EX. POLLICIS LONGUS
SERRATUS ANTERIOR
OCCIPITALIS PECTORALIS MAJOR
FRONTALIS
OUTER HAMSTRING TENDON
EX. DITITORUM COMMUNIS
OMOHYOID
PERONEUS TERTIUS TENDON
EX. CARPI RADIALIS LONGUS
BRACHIORADIALIS SOLEUS
RECTUS ABDOMINIS
SCALENUS MEDIUS
SEMISPINALIS CAPITIS
ABDUCTOR DIGITI MINIMI
RECTUS FEMORIS
SPLENIUS CAPITIS DELTOID
VASTUS LATERALIS TEMPORALIS
EX. DIGITORUM BREVIS
EX. DIGITORUM LONGUS BRACHIALIS
LEVATOR SCAPULAE TRICEPS BRACHII
SCALENUS ANTERIOR BICEPS BRACHII
GLUTEUS MAXIMUS ILIOTIBIAL BAND
TIBIALIS ANTERIOR BICEPS FEMORIS
PERONEUS LONGUS GASTROCNEMIUS
EX. POLLICIS BREVIS PERONEUS BREVIS
PERONEAL RETINACULUM

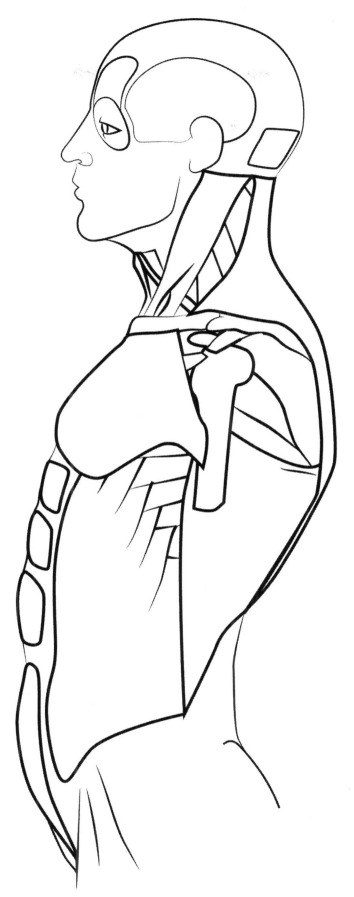

BODY LATERAL DEEP MUSCLES UNLABELED

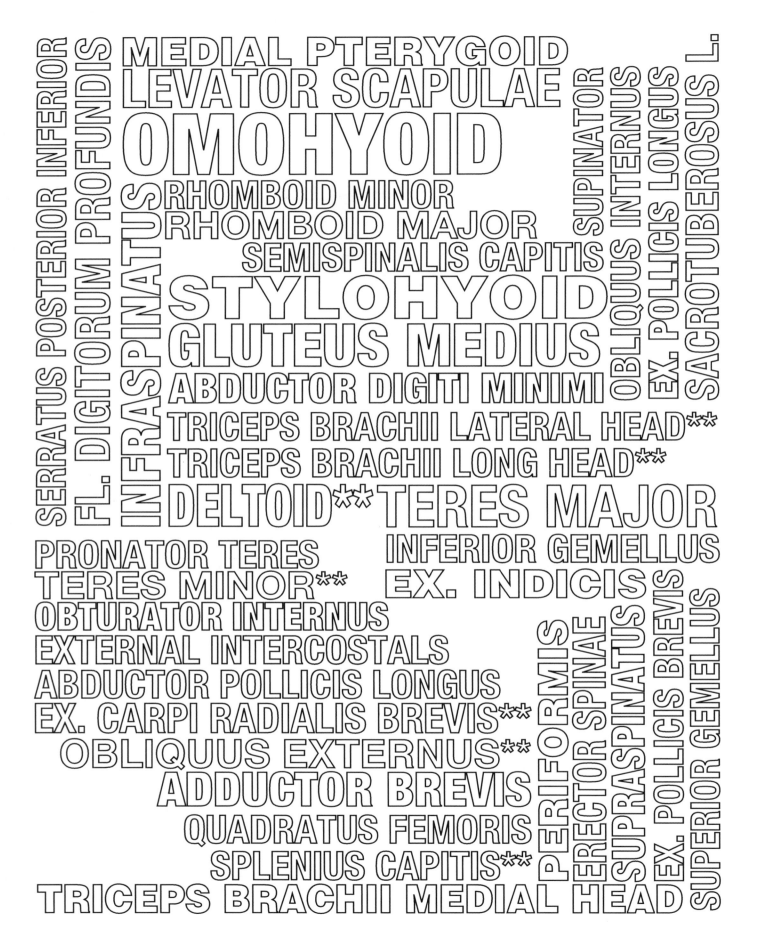

SERRATUS POSTERIOR INFERIOR
FL. DIGITORUM PROFUNDIS
INFRASPINATUS
MEDIAL PTERYGOID
LEVATOR SCAPULAE
OMOHYOID
RHOMBOID MINOR
RHOMBOID MAJOR
SEMISPINALIS CAPITIS
STYLOHYOID
GLUTEUS MEDIUS
ABDUCTOR DIGITI MINIMI
TRICEPS BRACHII LATERAL HEAD**
TRICEPS BRACHII LONG HEAD**
DELTOID**TERES MAJOR
SUPINATOR
OBLIQUUS INTERNUS
EX. POLLICIS LONGUS
SACROTUBEROSUS L.
PRONATOR TERES
TERES MINOR**
OBTURATOR INTERNUS
EXTERNAL INTERCOSTALS
ABDUCTOR POLLICIS LONGUS
EX. CARPI RADIALIS BREVIS**
OBLIQUUS EXTERNUS**
ADDUCTOR BREVIS
QUADRATUS FEMORIS
SPLENIUS CAPITIS**
INFERIOR GEMELLUS
EX. INDICIS
PERIFORMIS
ERECTOR SPINAE
SUPRASPINATUS
EX. POLLICIS BREVIS
SUPERIOR GEMELLUS
TRICEPS BRACHII MEDIAL HEAD

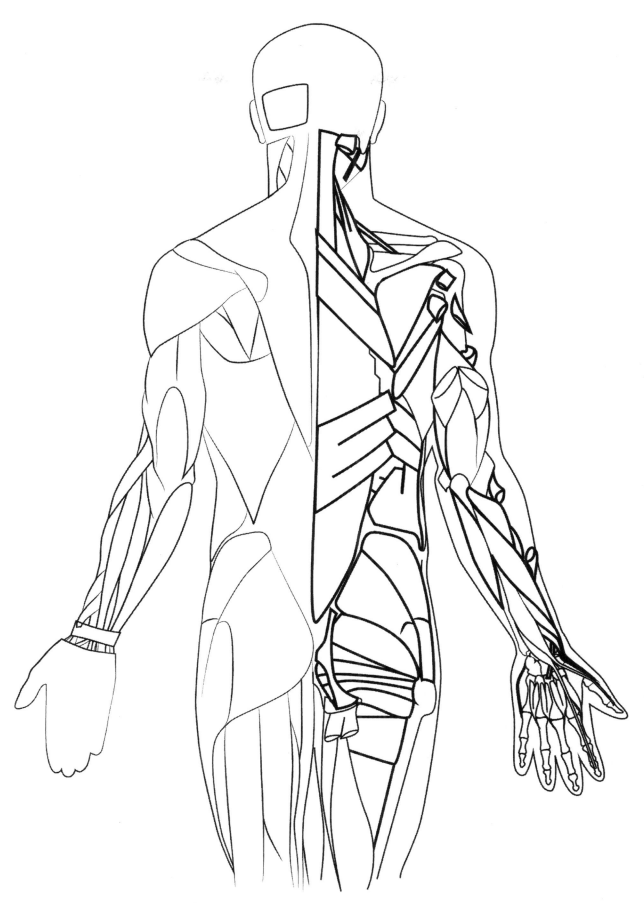

BODY POSTERIOR DEEP MUSCLES UNLABELED

PAGE | 23

Galea aponeurotica

Frontalis

Temporalis

Orbicularis oculi

Nasalis

Levator labii alaeque nasi

Levator labii superioris

Zygomatic minor

Zygomatic major

Buccinator

Orbicularis oris

Masseter

Mentalis

Depressor anguli oris

Corrugator

Levator palpebrae

Superior tarsus

Levator labii alaeque nasi**

Zygomatic minor**

Buccinator

Levator anguli oris

Muscular node

Masseter**

Mentalis**

HEAD ANTERIOR MUSCLES LABELED

NOTES

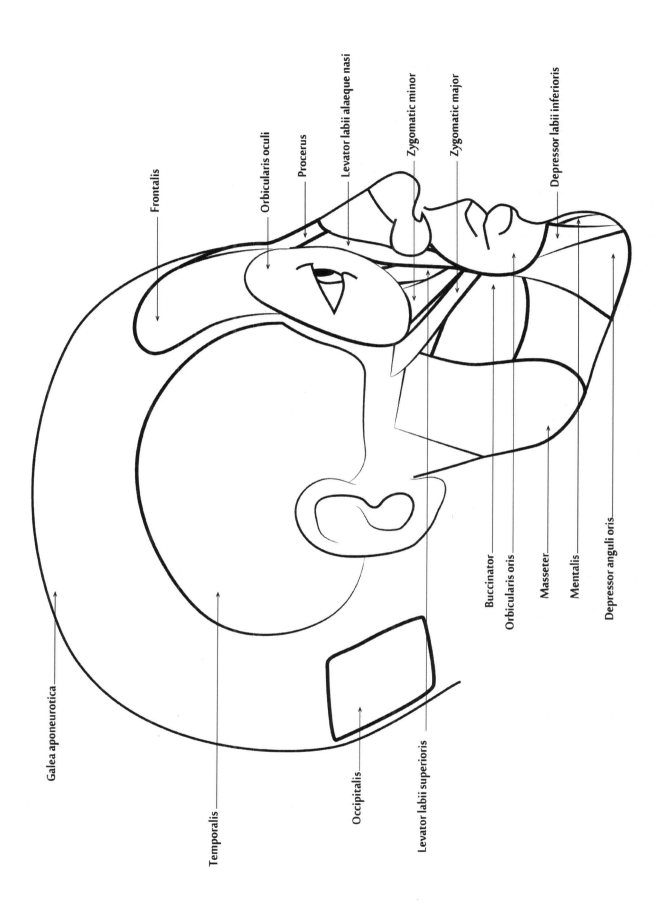

Galea aponeurotica

Frontalis

Orbicularis oculi

Procerus

Levator labii alaeque nasi

Zygomatic minor

Zygomatic major

Depressor labii inferioris

Temporalis

Occipitalis

Levator labii superioris

Buccinator

Orbicularis oris

Masseter

Mentalis

Depressor anguli oris

HEAD LATERAL MUSCLES LABELED

ZYGOMATIC MINOR**

GALEA APONEUROTICA

TEMPORALIS

MUSCULAR NODE

CORRUGATORS

LEVATOR LABII ALAEQUE NASI

FRONTALIS

SUPERIOR TARSUS

MASSETER**

MENTALIS**

BUCCINATOR

LEVATOR PALPEBRAC

ORBICULARIS ORIS

ZYGOMATIC MINOR

ZYGOMATIC MAJOR

BUCCINATOR

MASSETER

DEPRESSOR ANGULI ORIS

LEVATOR ANGULI ORIS

ORBICULARIS OCULI

NASALIS

LEVATOR LABII ALAEQUE NASI**

LEVATOR LABII SUPERIORIS

MENTALIS

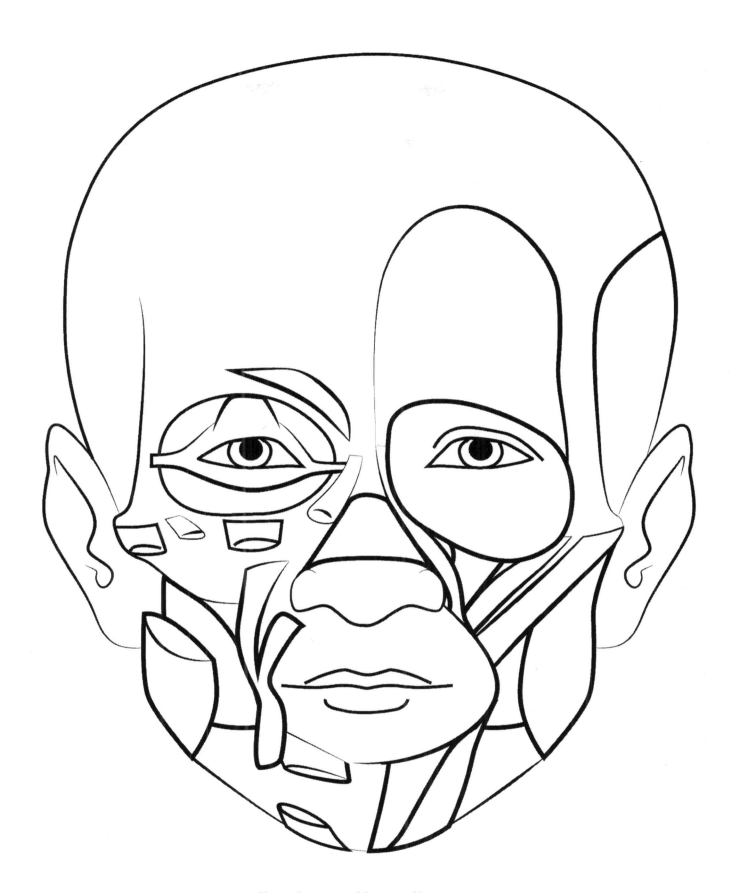

HEAD ANTERIOR MUSCLES UNLABELED

ZYGOMATIC MINOR

OCCIPITALIS

GALEA APONEUROTICA

TEMPORALIS

LEVATOR LABII SUPERIORUS

ORBICULARIS ORIS

MENTALIS

MASSETER

DEPRESSOR ANGULI ORIS

PROCERUS

BUCCINATOR

FRONTALIS

ORBICULARIS OCULI

LEVATOR LABII ALACQUE NASI**

DEPRESSOR LABII INFERIORIS

ZYGOMATIC MAJOR

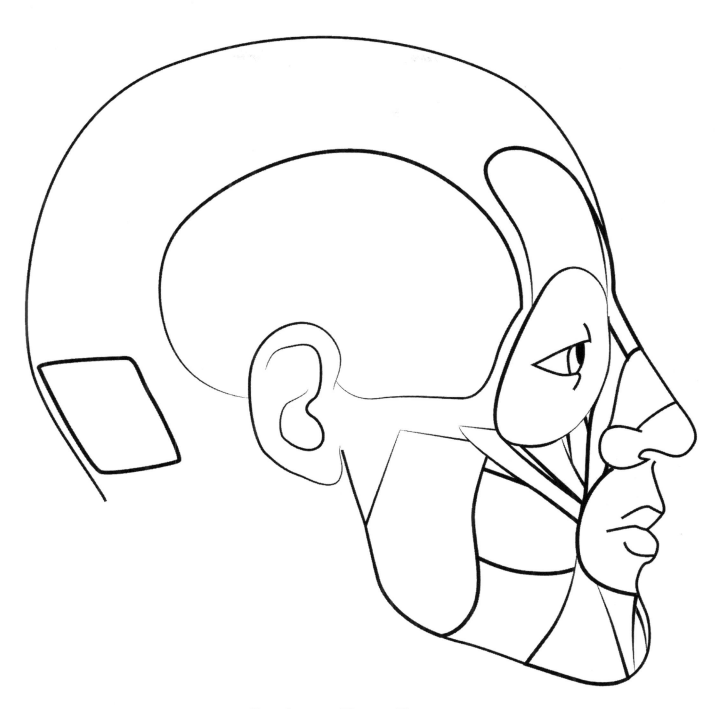

HEAD LATERAL MUSCLES UNLABELED

Notes

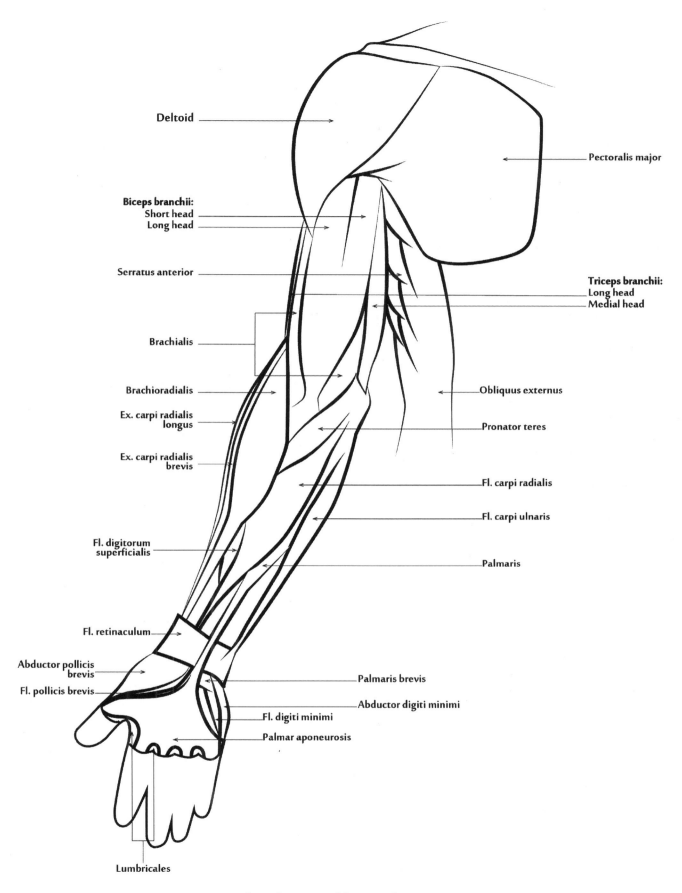

Deltoid

Biceps branchii:
Short head
Long head

Serratus anterior

Brachialis

Brachioradialis

Ex. carpi radialis
longus

Ex. carpi radialis
brevis

Fl. digitorum
superficialis

Fl. retinaculum

Abductor pollicis
brevis

Fl. pollicis brevis

Pectoralis major

Triceps branchii:
Long head
Medial head

Obliquus externus

Pronator teres

Fl. carpi radialis

Fl. carpi ulnaris

Palmaris

Palmaris brevis

Abductor digiti minimi

Fl. digiti minimi

Palmar aponeurosis

Lumbricales

ARM ANTERIOR MUSCLES LABELED

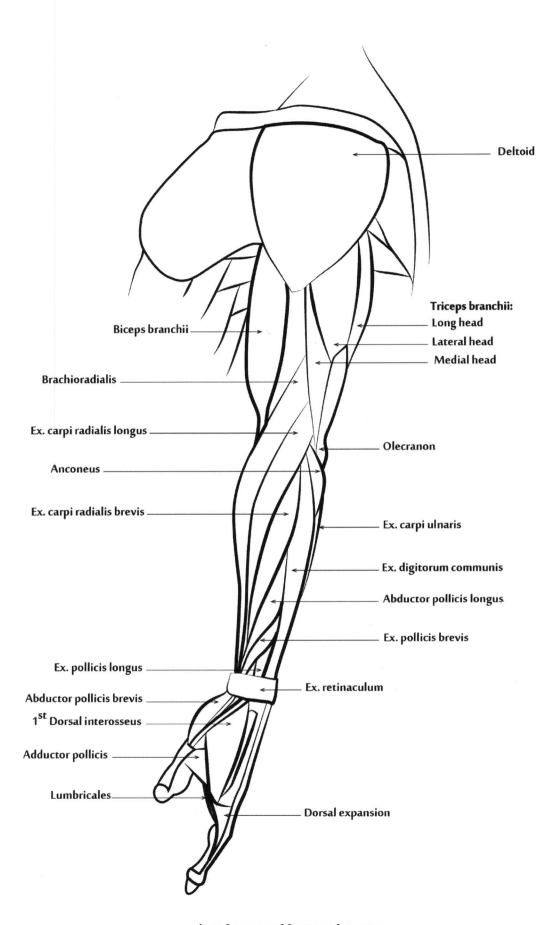

Deltoid

Triceps branchii:
Long head
Lateral head
Medial head

Biceps branchii

Brachioradialis

Ex. carpi radialis longus

Anconeus

Olecranon

Ex. carpi radialis brevis

Ex. carpi ulnaris

Ex. digitorum communis

Abductor pollicis longus

Ex. pollicis brevis

Ex. pollicis longus

Abductor pollicis brevis

1st Dorsal interosseus

Adductor pollicis

Ex. retinaculum

Lumbricales

Dorsal expansion

ARM LATERAL MUSCLES LABELED

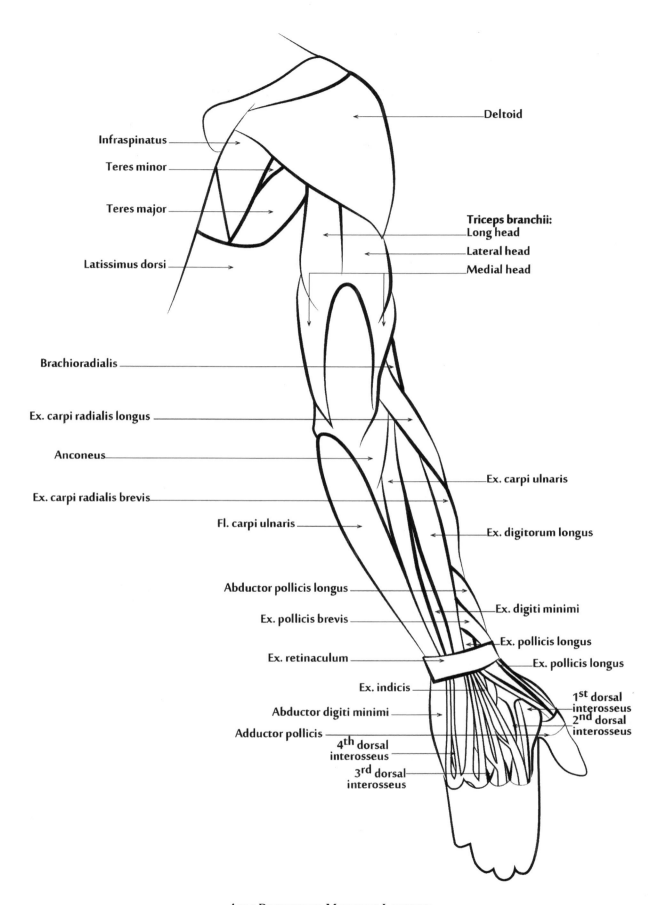

Infraspinatus

Teres minor

Teres major

Latissimus dorsi

Deltoid

Triceps branchii:
Long head

Lateral head

Medial head

Brachioradialis

Ex. carpi radialis longus

Anconeus

Ex. carpi radialis brevis

Fl. carpi ulnaris

Abductor pollicis longus

Ex. pollicis brevis

Ex. retinaculum

Ex. indicis

Abductor digiti minimi

Adductor pollicis

4th dorsal interosseus

3rd dorsal interosseus

Ex. carpi ulnaris

Ex. digitorum longus

Ex. digiti minimi

Ex. pollicis longus

Ex. pollicis longus

1st dorsal interosseus

2nd dorsal interosseus

ARM POSTERIOR MUSCLES LABELED

FL. DIGITI MINIMI PALMAR APONEUROSIS

SERRATUS ANTERIOVR

LONG HEAD BICEPS BRACHII:

ABDUCTOR DIGITI MINIMI

SHORT HEAD

DELTOID

PALMARIS

FL. RETINACULUM

PECTORALIS MAJOR

FL. POLLICIS BREVIS

FL. DIGITORUM SUPERFICIALIS

EX. CARPI RADIALIS BREVIS

PRONATOR TERES

OBLIQUUS EXTERNUS

EX. CARPI RADIALIS LONGUS

PALMARIS BREVIS

FL. CARPI ULNARIS

FL. CARPI RADIALIS

PALMARIS LONGUS

ABDUCTOR POLLICIS BREVIS

BRACHIALIS

LUMBRICALES

BRACHIORADIALIS

MEDIAL HEAD

TRICEPS BRACHII: LONG HEAD

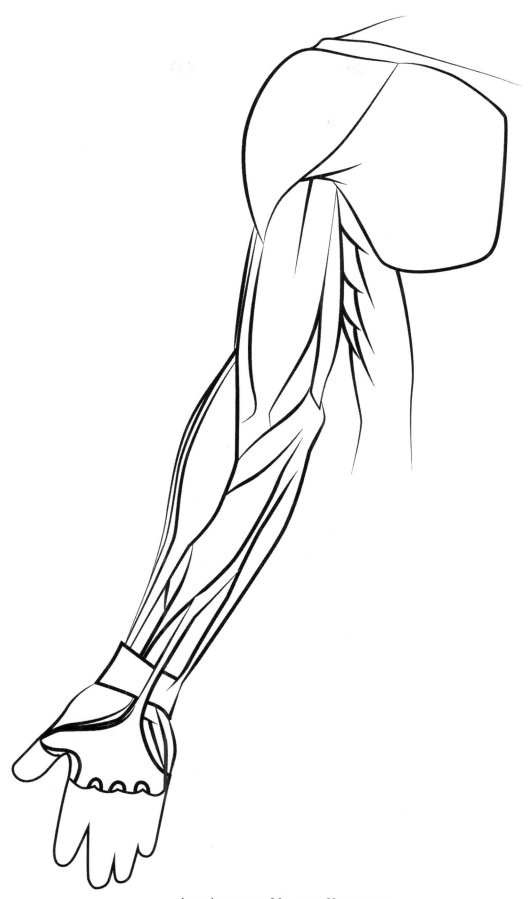

ARM ANTERIOR MUSCLES UNLABELED

ANCONEUS
1ST DORSAL INTERROSSEUS
EX. RETINACULUM
ADDUCTOR POLLICIS
TRICEPS BRACHII LATERAL HEAD
OLECRANON
EX. DIGITORUM COMMUNIS
ABDUCTOR POLLICIS LONGUS
EX. CARPI ULNARIS
BICEPS BRACHII
BRACHIORADIALIS
EX. POLLICIS LONGUS
DORSAL EXPANSION
DELTOID
LUMBRICALES
ABDUCTOR POLLICIS BREVIS
EX. POLLICIS BREVIS
EX. CARPI RADIALIS BREVIS
EX. CARPI RADIALIS LONGUS
TRICEPS BRACHII LONG HEAD
TRICEPS BRACHII MEDIAL HEAD

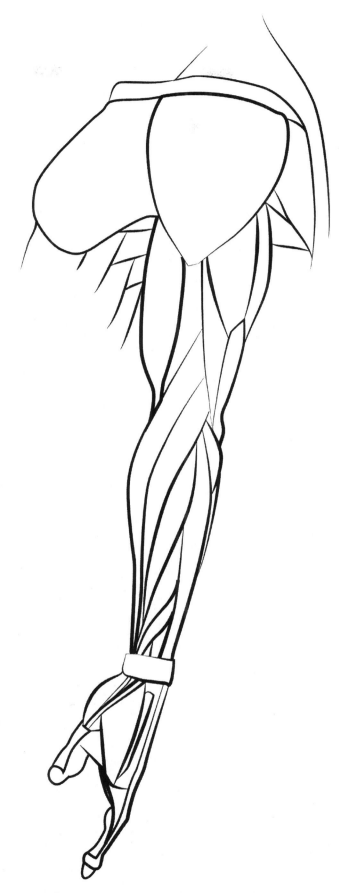

ARM LATERAL MUSCLES UNLABELED

INFRASPINATUS

FL. CARPI ULNARIS

ANCONEUS

EX. CARPI RADIALIS BREVIS

EX. CARPI RADIALIS LONGUS

ABDUCTOR POLLICIS LONGUS

TRICEPS BRACHII LONG HEAD

DELTOID

EX. DIGITI MINIMI

EX. RETINACULUM

EX. POLLICIS BREVIS

ADDUCTOR POLLICIS

ABDUCTOR DIGITI MINIMI

BRACHIORADIALIS

1ST DORSAL INTERROSSEUS

2ND DORSAL INTEROSSEUS

4TH DORSAL INTEROSSEUS

3RD DORSAL INTEROSSEUS

LATUSSIMUS DORSI

EX. POLLICIS LONGUS

MINOR

TERES

EX. CARPI ULNARIS

EX. DIGITORUM LONGUS

TRICEPS BRACHII MEDIAL HEAD

TRICEPS BRACHII LATERAL HEAD

TERES MAJOR

EX. INDICIS

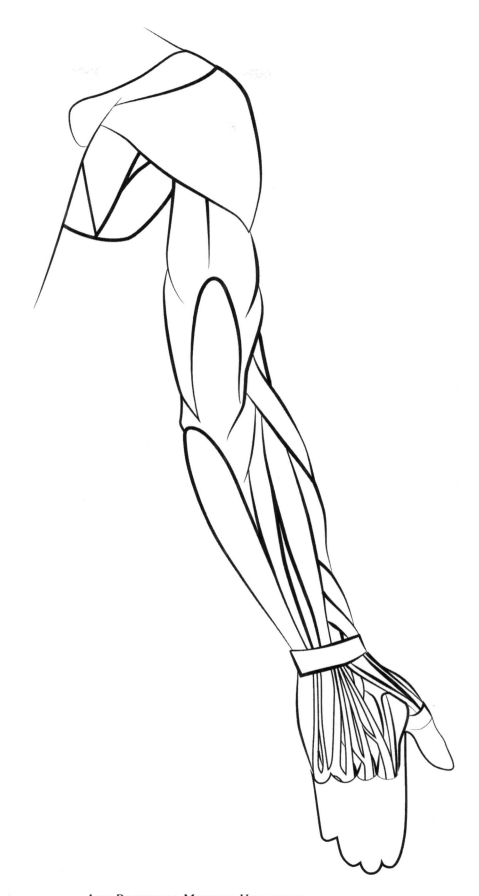

ARM POSTERIOR MUSCLES UNLABELED

NOTES

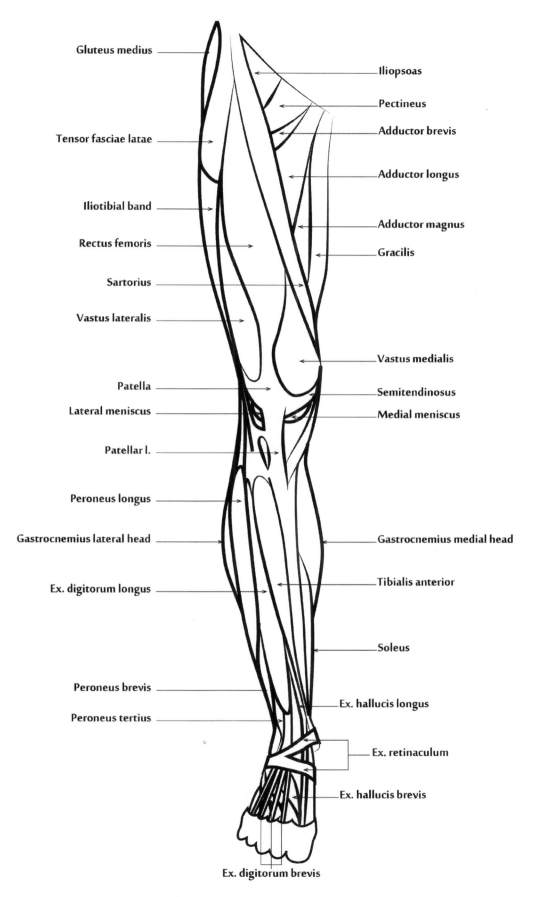

Gluteus medius

Iliopsoas

Pectineus

Adductor brevis

Tensor fasciae latae

Adductor longus

Iliotibial band

Adductor magnus

Rectus femoris

Gracilis

Sartorius

Vastus lateralis

Vastus medialis

Patella

Semitendinosus

Lateral meniscus

Medial meniscus

Patellar l.

Peroneus longus

Gastrocnemius lateral head

Gastrocnemius medial head

Ex. digitorum longus

Tibialis anterior

Soleus

Peroneus brevis

Peroneus tertius

Ex. hallucis longus

Ex. retinaculum

Ex. hallucis brevis

Ex. digitorum brevis

LEG ANTERIOR MUSCLES LABELED

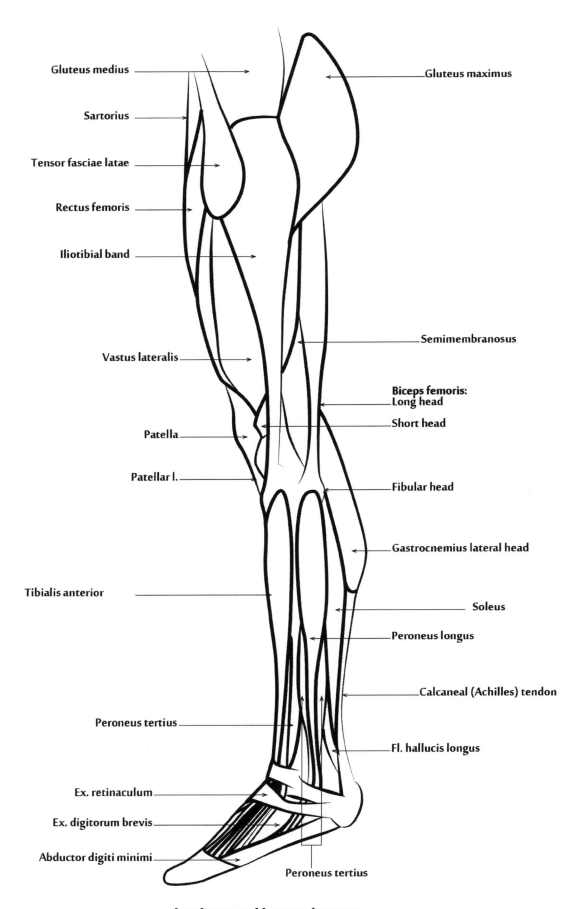

Gluteus medius

Sartorius

Tensor fasciae latae

Rectus femoris

Iliotibial band

Vastus lateralis

Patella

Patellar l.

Tibialis anterior

Peroneus tertius

Ex. retinaculum

Ex. digitorum brevis

Abductor digiti minimi

Gluteus maximus

Semimembranosus

Biceps femoris:
Long head

Short head

Fibular head

Gastrocnemius lateral head

Soleus

Peroneus longus

Calcaneal (Achilles) tendon

Fl. hallucis longus

Peroneus tertius

LEG LATERAL MUSCLES LABELED

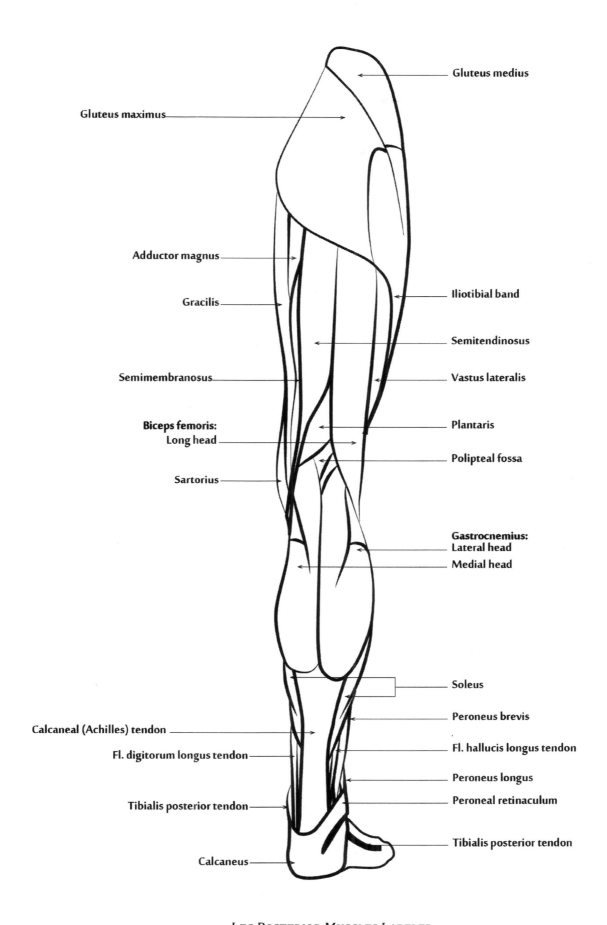

Gluteus medius

Gluteus maximus

Adductor magnus

Gracilis

Iliotibial band

Semitendinosus

Semimembranosus

Vastus lateralis

Biceps femoris:
Long head

Plantaris

Poplipteal fossa

Sartorius

Gastrocnemius:
Lateral head
Medial head

Soleus

Peroneus brevis

Calcaneal (Achilles) tendon

Fl. hallucis longus tendon

Fl. digitorum longus tendon

Peroneus longus

Tibialis posterior tendon

Peroneal retinaculum

Tibialis posterior tendon

Calcaneus

LEG POSTERIOR MUSCLES LABELED

PAGE | 45

PATELLAR L.
PATELLA
ILIOTIBAIL BAND
GRACILIS
EX. DIGITORUM LONGUS
SARTORIUS
VASTUS LATERALIS
ADDUCTOR BREVIS
PERONEUS BREVIS
SOLEUS
PERONEUS TERTIUS
EX. DIGITORUM BREVIS
TIBIALIS ANTERIOR
EX. HALLUCIS LONGUS

PECTINEUS
GLUTEUS MEDIUS
RECTUS FEMORIS
SEMITENDINOSUS
VASTUS MEDIALIS
MEDIAL MENISCUS
PERONEUS LONGUS
ADDUCTOR LONGUS
ADDUCTOR MAGNUS
LATERIAL MENISCUS
EX. HALLUCIS BREVIS
ILIOPSOAS
EX. RETINACULUM
TENSOR FASCIAE LATAE
GASTRONEMIUS LATERAL HEAD
GASTROCNEMIUS MEDIAL HEAD

LEG ANTERIOR MUSCLES UNLABELED

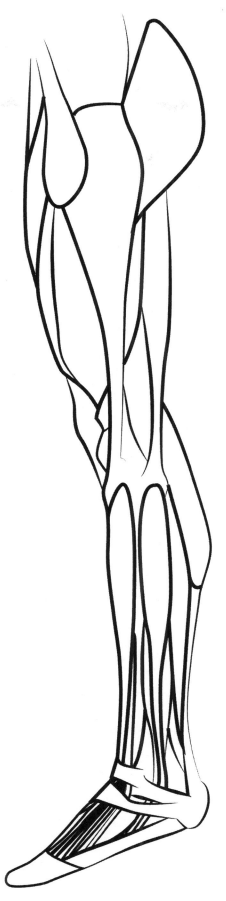

LEG LATERAL MUSCLES UNLABELED

PERONEUS LONGUS
SARTORIUS
CALCANEUS
GLUTEUS MEDIUS
SEMITENDINOSUS
GRACILIS
SEMIMEMBRANOSUS
BICEPS FEMORIS LONG HEAD
TIBIALIS POSTERIOR TENDON
SOLEUS
TIBIALIS POSTERIOR TENDON
GASTRONEMIUS: LATERAL HEAD
GASTROCNEMIUS MEDIAL HEAD
PLANTARIS
ILIOTIBAIL BAND
VASTUS LATERALIS
GLUTEUS MAXIMUS
POPLITEAL FOSSA
ADDUCTOR MAGNUS
PERONEUS BREVIS
PERONEAL RETINACULUM
FL. HALLUCIS LONGUS TENDON
CALCANEAL (ACHILLES) TENDON
FL. DIGITORUM LONGUS TENDON

Leg Posterior Muscles Unlabeled

PAGE | 51

NOTES

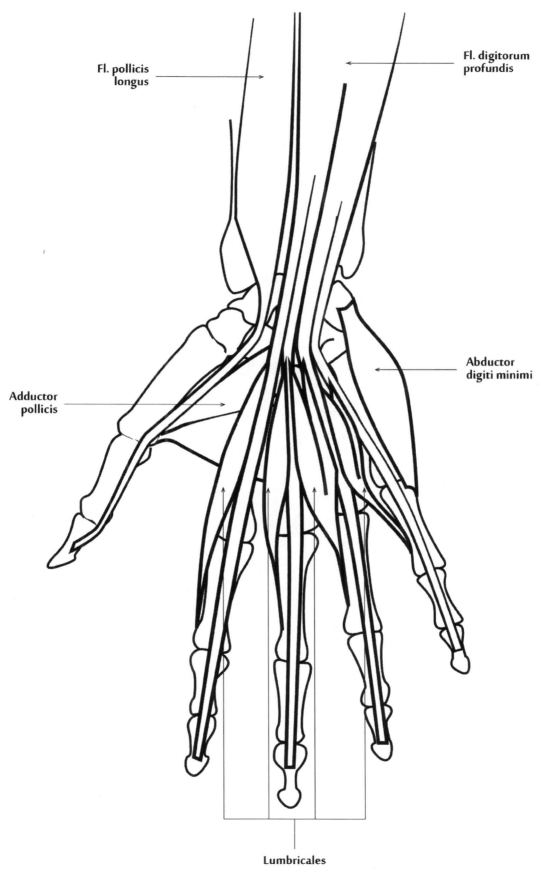

Fl. pollicis
longus

Fl. digitorum
profundis

Abductor
digiti minimi

Adductor
pollicis

Lumbricales

HAND PALMAR MUSCLES LABELED

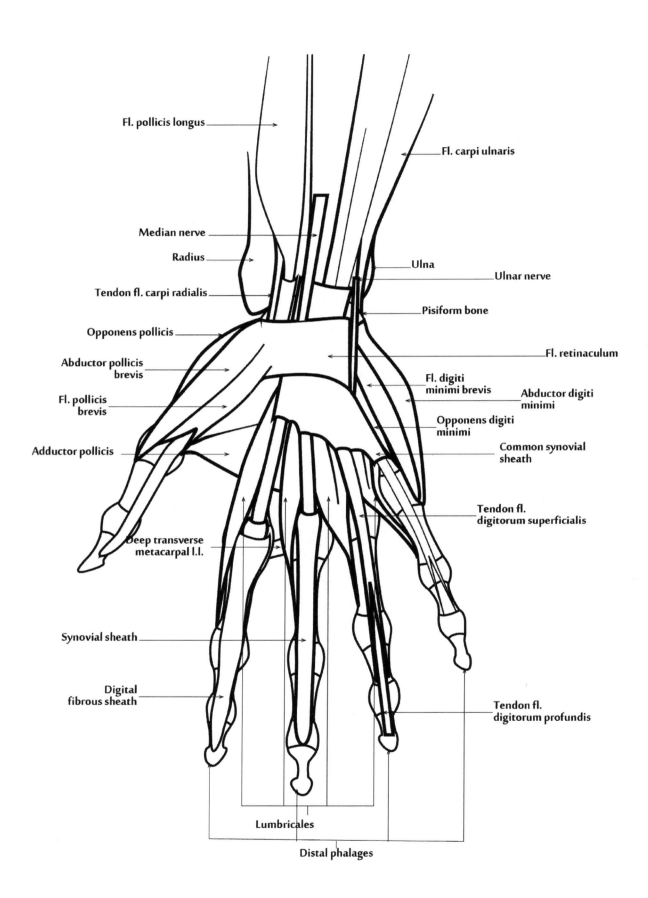

Fl. pollicis longus

Fl. carpi ulnaris

Median nerve

Radius

Ulna

Ulnar nerve

Tendon fl. carpi radialis

Pisiform bone

Opponens pollicis

Fl. retinaculum

Abductor pollicis
brevis

Fl. digiti
minimi brevis

Abductor digiti
minimi

Fl. pollicis
brevis

Opponens digiti
minimi

Adductor pollicis

Common synovial
sheath

Tendon fl.
digitorum superficialis

Deep transverse
metacarpal l.l.

Synovial sheath

Digital
fibrous sheath

Tendon fl.
digitorum profundis

Lumbricales

Distal phalages

HAND ANTERIOR MUSCLES LABELED

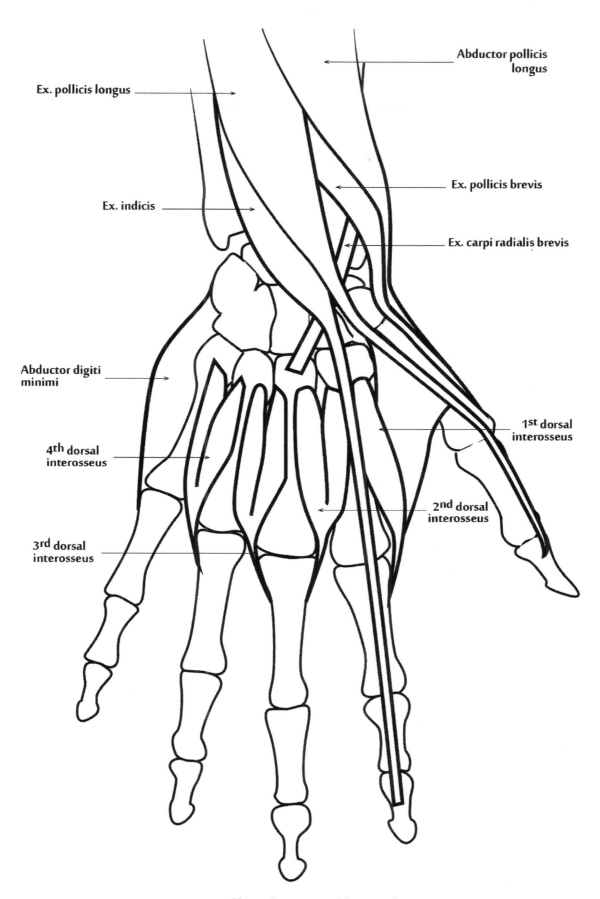

Abductor pollicis longus

Ex. pollicis longus

Ex. indicis

Ex. pollicis brevis

Ex. carpi radialis brevis

Abductor digiti minimi

1st dorsal interosseus

4th dorsal interosseus

2nd dorsal interosseus

3rd dorsal interosseus

HAND POSTERIOR MUSCLES LABELED

ADDUCTOR POLLICIS

ABDUCTOR DIGITI MINIMI

LUMBRICALES

FL. DIGITORUM PROFUNDIS

FL. POLLICIS LONGUS

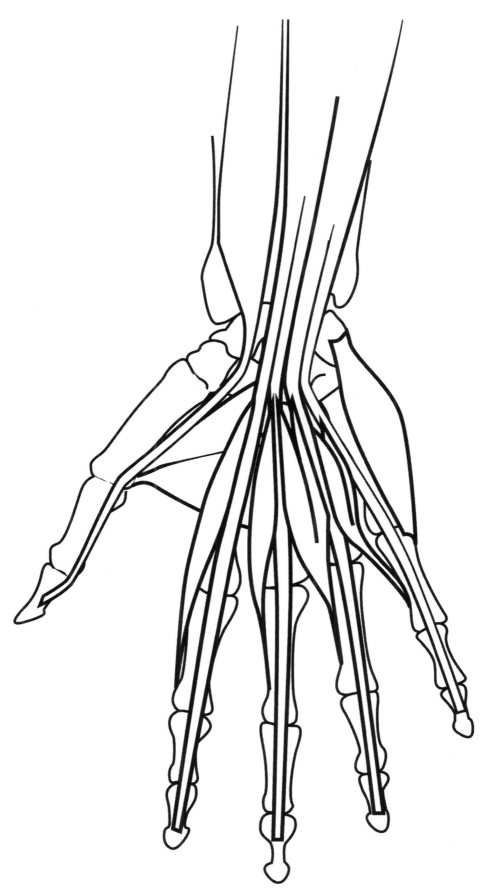

HAND PALMAR MUSCLES UNLABELED

DEEP TRANSVERSE METACARPAL L.L.

MEDIAN NERVE

RADIUS

OPPONENS POLLICIS

FL. CARPI ULNARIS

FL. DIGITORUM SUPERFICIALIS

FL. RETINACULUM

ABDUCTOR POLLICIS BREVIS

TENDON FL. CARPI RADIALIS

COMMON SYNOVIAL SHEATH

FL. DIGITORUM PROFUNDIS

DIGITAL FIBROUS SHEATH

FL. DIGITI MINIMI BREVIS

FL. POLLICIS BREVIS

FL. POLLICIS LONGUS

ADDUCTOR POLLICIS

LUMBRICALES

DISTAL PHALANGES

ULNA

ULNAR NERVE

TENDON FL. DIGITORUM SUPERFICIALIS

PISIFORM BONE

SYNOVIAL SHEATH

ABDUCTOR DIGITI MINIMI

OPPONENS DIGITI MINIMI

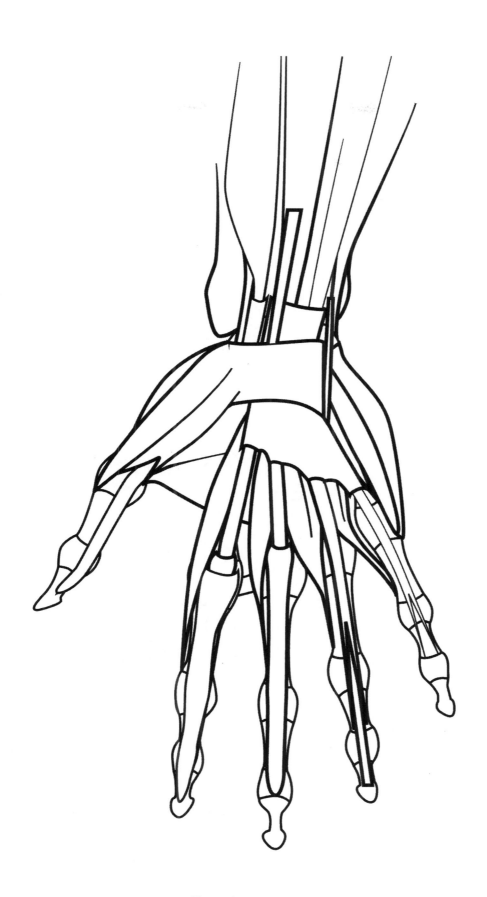

HAND ANTERIOR MUSCLES UNLABELED

2ND DORSAL INTEROSSEUS

4TH DORSAL INTEROSSEUS

ABDUCTOR DIGITI MINIMI

EX. POLLICIS LONGUS

EX. INDICIS

EX. CARPI RADIALIS BREVIS

EX. POLLICIS BREVIS

ADDUCTOR POLLICIS LONGUS

3RD DORSAL INTEROSSEUS

1ST DORSAL INTEROSSEUS

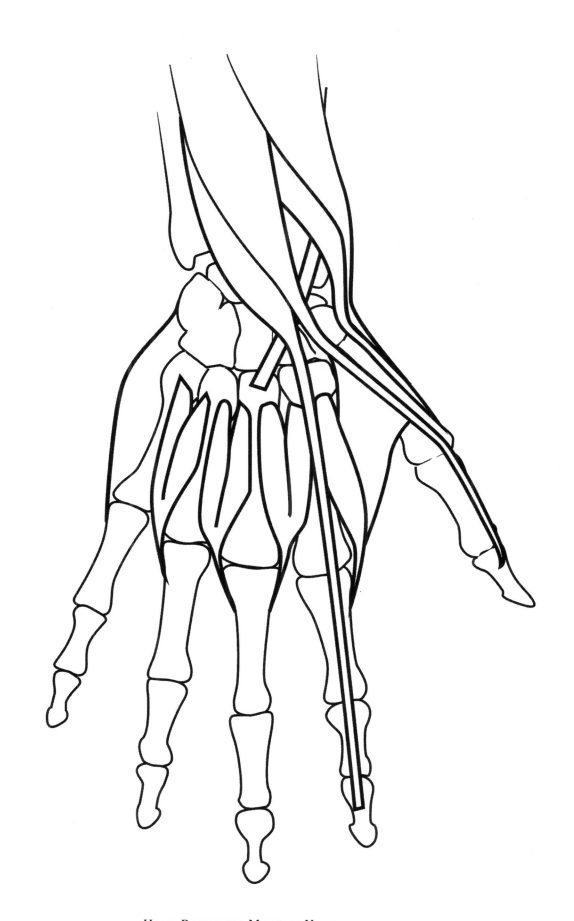

HAND POSTERIOR MUSCLES UNLABELED

NOTES

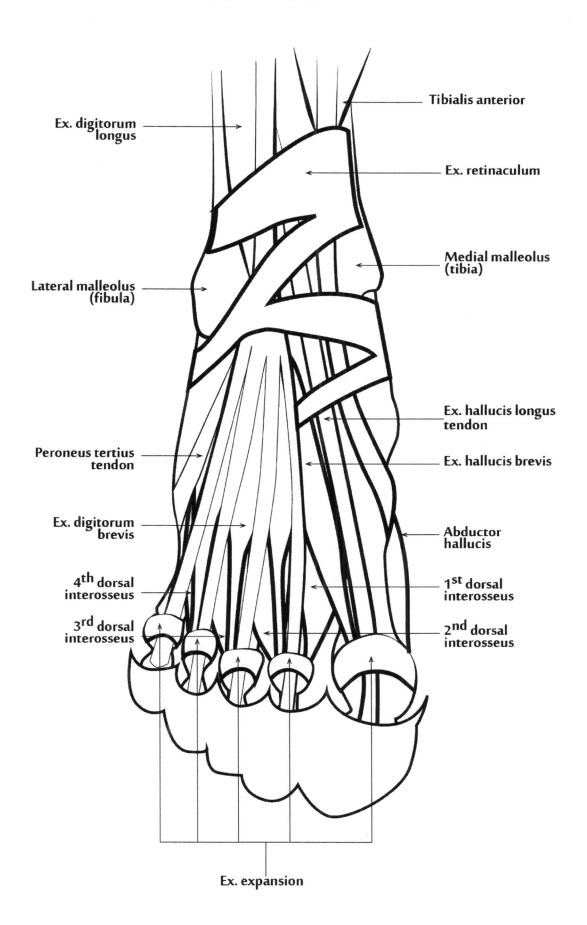

Ex. digitorum longus

Tibialis anterior

Ex. retinaculum

Medial malleolus (tibia)

Lateral malleolus (fibula)

Ex. hallucis longus tendon

Peroneus tertius tendon

Ex. hallucis brevis

Ex. digitorum brevis

Abductor hallucis

4th dorsal interosseus

1st dorsal interosseus

3rd dorsal interosseus

2nd dorsal interosseus

Ex. expansion

FOOT ANTERIOR MUSCLES LABELED

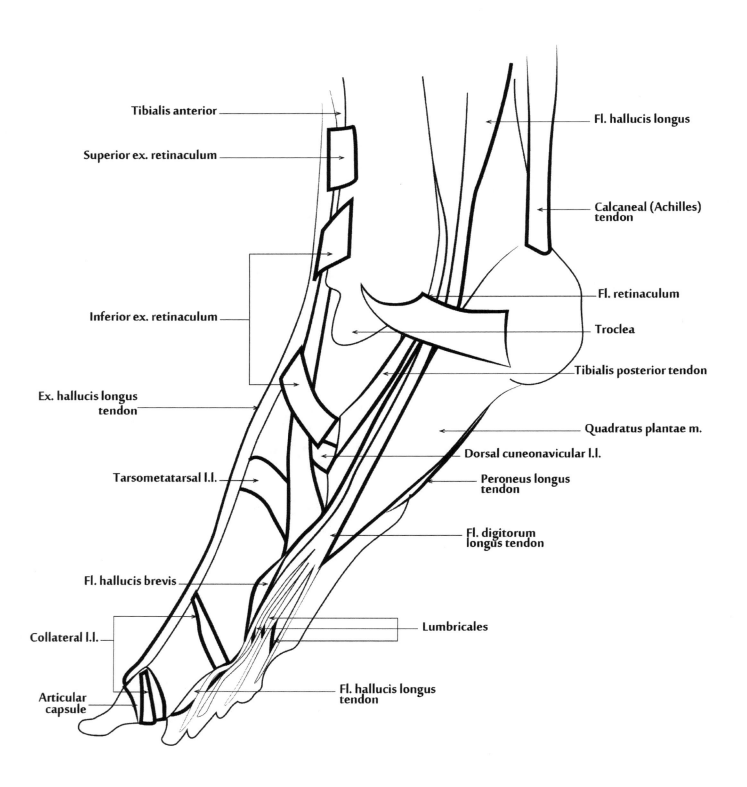

Tibialis anterior

Superior ex. retinaculum

Inferior ex. retinaculum

Ex. hallucis longus tendon

Tarsometatarsal l.l.

Fl. hallucis brevis

Collateral l.l.

Articular capsule

Fl. hallucis longus

Calcaneal (Achilles) tendon

Fl. retinaculum

Troclea

Tibialis posterior tendon

Quadratus plantae m.

Dorsal cuneonavicular l.l.

Peroneus longus tendon

Fl. digitorum longus tendon

Lumbricales

Fl. hallucis longus tendon

FOOT MEDIAL MUSCLES LABELED

PAGE | 64

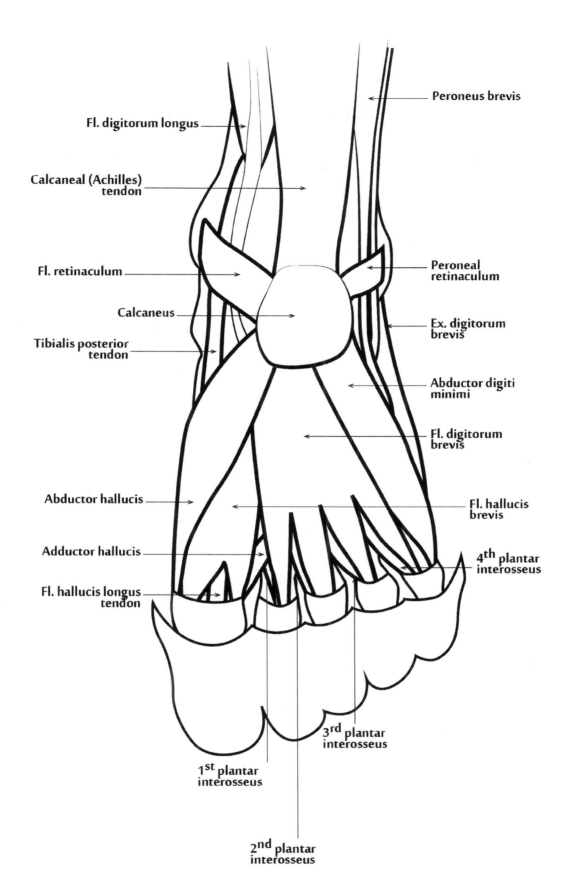

Fl. digitorum longus

Calcaneal (Achilles) tendon

Fl. retinaculum

Calcaneus

Tibialis posterior tendon

Abductor hallucis

Adductor hallucis

Fl. hallucis longus tendon

Peroneus brevis

Peroneal retinaculum

Ex. digitorum brevis

Abductor digiti minimi

Fl. digitorum brevis

Fl. hallucis brevis

4th plantar interosseus

3rd plantar interosseus

1st plantar interosseus

2nd plantar interosseus

FOOT PLANTER MUSCLES LABELED

MEDIAL MALLEOLS (TIBIA)

LATERAL MALLEOLUS (FIBULA)

EX. RETINACULUM

EX. HALLUCIS BREVIS

EX. EXPANSION

EX. DIGITORUM BREVIS

EX. DIGITORUM LONGUS

ABDUCTOR HALLUCIS

TIBIALIS ANTERIOR

1ST DORSAL INTEROSSEUS

2ND DORSAL INTEROSSEUS

3RD DORSAL INTEROSSEUS

4TH DORSAL INTEROSSEUS

PERONEUS TERTIUS TENDON

EX. HALLUCIS LONGUS TENDON

FOOT ANTERIOR MUSCLES UNLABELED

FL. HALLUCIS BREVIS
TIBIALIS ANTERIOR
COLLATERAL L.L.
INFERIOR EX. RETINACULUM
ARTICULAR CAPSULE
FL RETINICULUM
LUMBRICALES
TROCHLEA
SUPERIOR EX. RETINACULUM

FL. HALLUCIS LONGUS
TIBIALIS POSTERIOR TENDON
PERONEUS LONGUS TENDON
FT. HALLUCIS LONGUS TENDON
FL. HALLUCIS LONGUS TENDON
DORSAL CUNEONAVICULAR L.L.
TARSOMETATARSAL L.L.
QUADRATES PLATAE M.
FL. DIGITORUM LONGUS TENDON
CANCANEAL (ACHILLES) TENDON

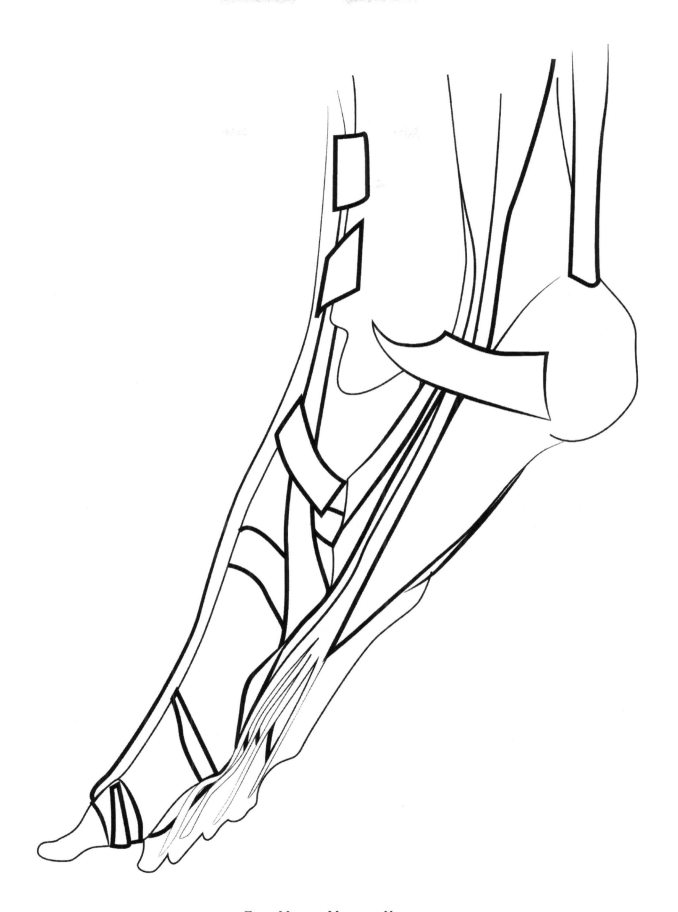

FOOT MEDIAL MUSCLES UNLABELED

PERONEUS BREVIS
ABDUCTOR HALLUCIS
ABDUCTOR DIGITI MINIMI
EX. DIGITORUM LONGUS

2ND PLANTAR INTEROSSEUS
1ST PLANTAR INTEROSSEUS
CANCANEAL (ACHILLES) TENDON

FL DIGITORUM LONGUS TENDON

CALCANEUS
FL RETINICULUM
PERONEAL RETINACULUM
EX. DIGITORUM BREVIS
FL. HALLUCIS BREVIS
3RD PLANTAR INTEROSSEUS
4TH PLANTAR INTEROSSEUS
FL DIGITORUM BREVIS
TIBIALIS POSTERIOR TENDON

FL. DIGITORUM LONGUS TENDON

FOOT PLANTER MUSCLES UNLABELED

About The Author

Our goal at Pamphlet books is to create content and illustrations that is very realistic, visually precise that communicate complex medical information that help educate medical students, medical professional and the general public.

We would love to know what you think about this coloring book and how to make it better. If you could take five minutes and go back to the site where you made your purchase and scroll to the bottom of that page and leave us a review.